621.042023 W85o 1992
Woodburn, John H.
Opportunities in energy
 careers

OPPORTUNITIES IN
ENERGY CAREERS

VGM Opportunities Series

OPPORTUNITIES IN
ENERGY CAREERS

John H. Woodburn

VGM Career Horizons
a division of *NTC Publishing Group*
Lincolnwood, Illinois USA

Cover Photo Credits:
Front cover: upper left and lower right, UOP; upper right and lower left, American Petroleum Institute.

Back cover: upper left and lower right, UOP; upper right and lower left, American Petroleum Institute.

Allen County Public Library
900 Webster Street
PO Box 2270
Fort Wayne, IN 46801-2270

Library of Congress Cataloging-in-Publication Data

Woodburn, John H.
   Opportunities in Energy careers / John H. Woodburn.
      p. cm. — (VGM opportunities series)
   ISBN 0-8442-4009-5 — ISBN 0-8442-4010-9 (pbk.)
     1. Power resources—Vocational guidance. I. Title. II. Series.
TJ163.2.W64 1992
621.042'023—dc20   91-39559
                                                                         CIP

Published by VGM Career Horizons, a division of NTC Publishing Group.
© 1992 by NTC Publishing Group, 4255 West Touhy Avenue,
Lincolnwood (Chicago), Illinois 60646-1975 U.S.A.
All rights reserved. No part of this book may be reproduced, stored
in a retrieval system, or transmitted in any form or by any means,
electronic, mechanical, photocopying, recording or otherwise, without
the prior permission of NTC Publishing Group.
Manufactured in the United States of America.

2 3 4 5 6 7 8 9 0 VP 9 8 7 6 5 4 3 2 1

ABOUT THE AUTHOR

Dr. John H. Woodburn has been teaching in the Montgomery County, Maryland, public schools since 1960, after holding teaching and administrative positions at The Johns Hopkins University, the National Science Teachers Association, Illinois State University, and Michigan State University. He also held consultant appointments with the U.S. Office of Education and the Atomic Energy Commission.

Books he has written for young people reveal his concern for the intellectual and personal capabilities that appear to be related to the successful pursuit of science-related careers. Titles of his books include: *Excursions into Chemistry, The Whole Earth Energy Crisis,* and *Taking Things Apart and Putting Things Together.* He is also the author of another VGM Career Horizons title, *Opportunities in Chemistry.* Dr. Woodburn holds an A.B. degree from Marietta College, an M.A. from Ohio State University, and a Ph.D. from Michigan State University.

CONTENTS

About the Author v

1. **Introduction to Energy Careers** 1
   Distribution and utilization of earth's energy resources. Energy and career opportunities. Three examples of energy careers. Energy and the population problem. The energy supply and our life-styles. Energy awareness.

2. **Energy: What It Is and What It Does** 21
   Defining energy. Energy measuring units and how to exchange them. How scientists describe energy. Energy and the environment.

3. **Careers in the Oil Industry** 31
   The nature of this energy resource. History of the petroleum industry. Kinds and numbers of job opportunities. Education and training. Opportunities for women. Trends in employment. What about oil shale and tar sands? Oil industry jobs at a glance.

4. **Careers in the Coal Industry** 59
   The nature of this energy resource. History of the coal industry. Types of coal mines. Mining processes. Kinds and numbers of job opportunities. Education and train-

ing. Trends in employment. Coal and environmental pollution. Coal industry jobs at a glance.

5. **Careers in the Natural Gas Industry** 73

   The history of the gas industry. Estimating the world supply of energy resources. Career opportunities in the gas industry.

6. **Careers in the Solar Energy Industry** 79

   Understanding solar energy. The nature of this energy resource. The solar energy industry. From sunlight to electricity. Solar energy jobs at a glance.

7. **Careers in the Nuclear Energy Industry** 93

   History of the nuclear energy industry. Trends in employment. The safety and sanity of nuclear reactors. Nuclear energy jobs at a glance.

8. **Careers in the Electric Power Industry** 111

   History of the electric power industry. Applying for a job. Some job descriptions. Trends in employment. Power plant jobs at a glance.

9. **Energy from the Earth's Nondepletable Sources** . . 125

   Wind power. Waterpower. Energy from oceans. Energy from osmosis and photosynthesis. Biofuels as a source of energy.

10. **Energy Careers: Some Common Themes** 147

    Sociopolitical implications. Scientific infrastructure. Decision making.

CHAPTER 1

INTRODUCTION TO ENERGY CAREERS

Highly rewarding careers for thousands of men and women exist in producing, distributing, and conserving energy. Furthermore, energy has begun to assume a leading role in determining how well people throughout the global community will be able to live.

People everywhere, but particularly in the less-industrialized nations, are striving to improve their ways of living. They want to reduce the misery of hunger and inadequate housing, to improve communication and transportation facilities, to combat disease, to give children a better education, to have greater freedom to express creativity—in short, to achieve the quality of life that has long been the goal of people in the highly industrialized nations. But these ambitions can be fulfilled only by increasing the per capita consumption of energy.

DISTRIBUTION AND UTILIZATION OF EARTH'S ENERGY RESOURCES

Problems arise because the earth's energy resources are not distributed equally among all nations. This is particularly true of the store of the so-called fossil fuels: coal, oil, and gas. Increasing demands on fundamentally finite supplies are sure to heighten competition and thereby give energy an even greater role in maintaining good relations among nations. The responsibility for wise decisions regarding the development, management, and conservation of energy falls on all citizens and their leaders. Helping them to fulfill this responsibility adds to the appeal of energy-related careers.

A couple of examples will demonstrate how unequally the earth's energy resources are distributed among world populations. The Persian Gulf countries, which represent scarcely 2 percent of the total world population, control an estimated one-half of the earth's petroleum. On the other hand, more than a third of the world's people live in China and India, but these nations control only an estimated 3 percent of the world supply of oil.

Nations also differ widely in how rapidly they consume their energy resources. For example, China and India have low per capita energy utilization. Were these large populations to strive to make more energy available en route to improved life-styles, there would be greater competition among all nations.

It is becoming increasingly evident that careers involving the development, distribution, utilization, and conservation of energy will be closely linked with the sociopolitical solutions to terribly crucial problems. For example, will nations that have endured low living standards be willing to share their not-yet-developed energy resources with those nations that have enjoyed high living standards but are running out of readily available energy supplies? Although all the sociopolitical circumstances that led up to the Persian Gulf War of 1990-1991 are not yet clear, the availability of energy resources undoubtedly was a major factor.

Just as energy plays a vital role in everything people want and need to do, it is equally involved in everything that happens in our natural environment. Weather and climate and their effects on the biosphere provide the most obvious evidence of energy's role in our surroundings. The total effects are far-reaching, as the example of the so-called greenhouse warming of the earth shows. The more we need to be concerned about the quality of the environment, the more we must take into account the availability, utilization, and conservation of the world's energy resources. This close linkage between environmental quality and energy creates an expanding need for people who are

Table 1.1: World Recoverable Energy Resources

OIL IN BILLIONS OF BARRELS

Country	Amount Remaining in Proved Reserves (1985)	Annual Consumption (1984)	R/P^1	Identified Plus Probable Reserves	Total R/P^2
United States	47.4	3.25	14.6	220.0	67.7
Canada	6.0	.53	11.3	44.7	84.3
Mexico	30.0	.98	30.5	66.9	68.2
South America	58.6	1.32	44.5	138.0	104.5
Africa	68.5	1.61	42.7	140.9	87.6
Middle East	420.9	4.20	100.3	601.1	157.4
U.S.S.R.	81.0	4.48	18.1	247.2	55.2
China	23.6	.84	34.0	69.3	82.5
India	6.2	.20	31.3	10.3	57.5
Australia and New Zealand	2.0	.19	10.6	8.2	43.2
Europe	35.2	1.43	24.2	72.7	50.8

NATURAL GAS IN TRILLIONS OF CUBIC FEET

Country	Amount Remaining in Proved Reserves (1985)	Annual Consumption (1984)	R/P^1	Identified Plus Probable Reserves	Total R/P^2
United States	326.9	18.08	18	1898.8	105.0
Canada	99.2	3.59	25	483.2	134.6
Mexico	76.7	1.43	11	211.4	147.8
South America	123.5	1.90	64	292.9	154.1
Africa	234.7	1.87	125	582.2	311.3
Middle East	1186.9	1.32	900	2149.0	1628.0
U.S.S.R.	1275.4	19.40	65	2819.3	145.3
China	24.8	0.37	67	234.3	633.2
India	19.2	0.15	128	36.6	244.0
Australia and New Zealand	72.9	0.59	124	137.0	232.0
Europe	297.4	8.35	35	517.2	61.9

[1] Amount remaining in proved reserves divided by the 1985 rate of annual consumption.
[2] Amount remaining in proved reserves plus the amount in probable reserves divided by the 1985 rate of annual consumption.

Adapted from: "World Estimates of Original Reserves and Ultimate Recoverable Resources of Conventional Crude Oil . . . and Conventional Natural Gas," Masters, Charles D., Emil D. Attanasi, William D. Dietzman, Richard F. Meyer, Robert W. Mitchell, and David H. Root. *12th World Petroleum Congress, 1987, Proceedings, Vol. 5, pp. 3–27.*

Table 1.2: Per Capita Energy Utilization and Related Factors in Selected Industrialized and Yet-to-be Industrialized Nations

Nation	Per Capita Annual Energy Use in Millions of BTUs	Life Expectancy	Birth Rate Per Thousand
United States	324	76	14
Canada	408	76	12
Mexico	48	70	26
Brazil	36	66	26
United Kingdom	169	76	13
France	170	76	13
Switzerland	160	77	11
U.S.S.R.	208	70	16
Egypt	25	63	30
Saudi Arabia	22	65	41
India	8	60	31
China	23	70	19
Japan	142	78	12
World Total	63	63	26

Adapted from: *International Energy Annual, 1989*, U.S. Department of Energy, and *The Universal Almanac*, 1991. Edited by John W. Wright. Kansas City: Andrews and McMeel.

ready, willing, and able to deal simultaneously with scientific, technological, economic, and sociopolitical problems.

ENERGY AND CAREER OPPORTUNITIES

What is being said here applies to a wide array of career options—options that are as nearby as tending service stations, as necessary as restoring electrical power in storm-ravaged communities, as traditional as gathering firewood or mining coal, as imaginative as building floating factories to harvest the solar energy that is trapped in the surface waters of tropical oceans, or as high-tech as designing satellites to collect solar energy in space and deliver it to wherever it is needed.

Production and Distribution

The production and distribution of energy provides jobs for secretaries, engineers, accountants, personnel specialists, mathematicians, physicists, custodial and maintenance workers, and people in dozens of other trades and professions. And there is something about working in an energy-related activity that ties these jobs together with a unique sense of importance. It is a subtle awareness of contributing to the continued well-being not only of one's immediate community but also of the nation.

Energy-related events and issues are increasingly becoming featured news items—news that touches on the entire citizenry's well-being. For example, early in the 1970s, people in the United States were shocked into becoming painfully aware of how dependent they were on their neighborhood service stations. Prior to those terribly confusing days of gasoline shortages and empty fuel tanks, people took for granted the seemingly endless supply of the fossil fuels. Arguments still go on about what actually caused gasoline stations to run out of gas and fuel suppliers to ration their product during the winter of 1973. But this one event became a well-taught lesson.

In 1989, the oil tanker *Exxon Valdez* ran aground and spilled barrels of oil into Alaska's Prince William Sound. That the spill resulted in the loss of enough oil to supply the United States for approximately forty-five minutes received much less emphasis in the news media than did the impact of the spill on the environment, particularly the waterfront near the town of Valdez. For weeks the evening news featured oil-soaked waterfowl and sea otters suffering and dying. In what seemed to be a totally futile effort, thousands of hours were spent literally wiping oil from millions of pebbles.

The oil company that was responsible for this environmental catastrophe lost approximately $3 billion. This figure included $1 billion to compensate the Alaskan people for loss of income from sources that were temporarily unavailable and for the expenses involved in restoring their environment to a livable condition. It is impossible, however, to put a dollar value on the aesthetic and psychological damage resulting from having one's environment and the total biosphere become polluted with oil.

When viewed in retrospect, the Valdez incident emphasizes the challenges that await people who seek careers that involve the interdependence between insuring adequate supplies of energy and maintaining the quality of the environment. Not yet do we know how best to take advantage of nature's ways of returning petroleum to its original nonpolluting elements. It is this kind of knowledge that stands the best chance of enabling us to head off or solve the environmental problems that occur all along the way between utilizing nature's stores of energy and returning the waste products to their natural origins.

Efficiency Improvement

Careers devoted to improving the efficiency of energy utilization are already making valuable contributions. Already, new lamps are available that are 90 percent more efficient than previous ones. Newly designed household appliances require 20 percent less electricity than earlier models. New automobiles that yield 60 miles per gallon of fuel are already available. Builders of new homes can guarantee their owners that heating bills will not exceed $200 annually if they install hot water heaters that double as furnaces.

The possibility of developing products or techniques that combine improved efficiency with protection of the environment makes careers in energy increasingly worthwhile. For example, efficiency-oriented research has reduced the amount of energy required to produce polyethylene by more than 75 percent since its invention in 1940. Recycling this widely used plastic promises not only to increase the efficiency of energy utilization, but also to reduce the amount of energy consumed in disposing of this solid waste. In general, recycling bypasses the most energy-consuming steps in manufacturing. The aluminum industry, for example, tells us that it takes only half as much energy to recycle aluminum as it does to recover aluminum from its source in nature.

The satisfactions that come with careers in finding, developing, and distributing the earth's energy resources are well known. These satisfactions will surely continue, but they are becoming increasingly elusive as new deposits become more difficult to find or harvest. Increased competition will call for people who know more and more about how the earth's energy resources were created, where they are most likely to occur, and how they can be developed more efficiently. Transporting these resources from where they occur to where they are needed also creates thousands of career opportunities—opportunities that are particularly challenging because transport accidents pose enormous environmental threats.

If the fossil fuels become scarce, economic and sociopolitical problems will become critical. It will become increasingly urgent to develop renewable energy resources. All of nature's ways to capture, store, and redistribute solar energy will attract more attention. There also will be more emphasis on improving the safety and efficiency of nuclear reactors. Of particular concern are problems in disposing of radioactive wastes from nuclear fission and controlling the fantastic amount of energy that accompanies nuclear fusion. All of these considerations will have an obvious impact on future career opportunities.

An awesome array of career opportunities also stems from energy's role in the relationships and responsibilities shared among nations. The circumstances leading up to and resulting from the Persian Gulf War of 1990-91 present a case in point. The link between the development of nuclear energy and the possible spread of its use in weaponry adds another dimension to energy's role in international relations.

Society's energy-related problems generate controversy, however. Personal interests and concerns cause people to view these problems differently and, consequently, to differ on what may be the best solutions. Differences can create obstacles. In fact, the more a problem affects the lives of citizens, the greater must be the knowledge and forbearance of those who respond to the challenges posed by energy-related careers.

THREE EXAMPLES OF ENERGY CAREERS

What we are saying is documented by three examples of the thousands of men and women who have responded to these challenges. The careers of these three individuals take us from the ocean depths, into the biosphere, and to outer space. They combine interests and competence in science, technology, government, politics, ethics—in fact, in nearly all human endeavors.

Peter E. Glaser, vice-president of Arthur D. Little Corporation Inc., has devoted much of his professional career to designing satellites that will collect solar energy in space and transmit it to earth. William H. Avery is the former director of the Ocean Energy Programs at the Johns Hopkins University Applied Physics Laboratory. His energy-related projects include designing a system for collecting and distributing the solar energy that is stored in the surface waters of tropical oceans. Sherwood B. Idso, a research physicist with the USDA's Agricultural Research Service, has followed a career that combines making energy available and, at the same time, taking into account how the utilization of energy possesses the capacity to change the entire face of the planet and the circumstances of all who inhabit it.

Glimpses of the life stories of these three people tell us much about what it takes to prepare for and to gain the satisfactions that derive from energy-related careers. By occasionally reading between the lines, we can also sense the personality traits that best overcome the obstacles that get in the way when people tackle energy-related problems and issues. As seen by Idso:

> There should be little doubt in the mind of anyone that the times in which we live are indeed unique; for in a development never before experienced in the history of life on Earth, a single species has emerged from the backdrop of the biosphere to gain such power and prominence that it possesses the capacity to totally change the face of the planet and the circumstances of all that inhabit it. The species, of course, is man; and the phenomenon, his unbridled passion for processing the planetary resources upon which all of life depends for sustenance

and for returning the remains to the environment as the altered waste products of his supra-metabolic activity.

Foremost among the host of societal effluents thus foisted upon the fabric of nature is carbon dioxide or CO_2, expired directly by all animal life, but produced by man most prodigiously through the burning of fossil fuels and the felling of forests. So significant have these latter activities become over the past few centuries, in fact, that man is currently responsible for increasing the concentration of this trace atmospheric constituent by a full 30 percent, with half of the increase occurring in just the past four decades. (*Carbon Dioxide and Global Change: Earth in Transition:* IBR Press, 1989.)

Avery is a good example of those whose careers include many interests. One of his interests concerns extending the world's supply of available energy. He is well trained in chemistry and physics, is compatible with technology, undaunted by government and bureaucracy, understands the point of view of business and industry, and is sensitive to public concern over the possible impact of energy resources that are being developed. Avery sees "the solar energy absorbed every day in the tropical oceans as equivalent to the heat that would be produced by burning 160 billion barrels of oil per day. This is four thousand times the daily world oil production in 1980. A practical and economic method for using even a tiny portion of this inexhaustible energy could have far-reaching benefits to the U.S. and the rest of the world. Ocean thermal conversion offers the possibility." (*Ocean Thermal Energy Conversion:* Oxford University Press, 1991.) Avery shows that factories on ships cruising in the tropical oceans could use the ocean thermal energy conversion (OTEC) process to produce enough methanol from seawater and coal to free the United States from dependence on imported oil.

Approaching the same problem but from a different point of view, Glaser reminds us that technological advances have been happening at a dizzying pace during the twentieth century. Outer space is being explored, the forces within atoms are now sources of energy, recombinant DNA strategies promise to bring the characteristics of plants and animals within humanity's grasp, and methods of electronic communication are creating a "global village." At the same time, "the illusion that man has unlimited capabilities to control nature and fashion the environment based on scientific understanding, and technological prowess has engendered a naive belief that man can control nature and exploit energy and material resources with impunity to meet his immediate needs. Only scant attention is paid to the reality that these resources are irreplaceable assets and that their profligate use may threaten the global environment and even the conditions under which future generations will have to live." ("Power from Space for Use on Earth: An Emerging Global Option," from *Space Commercialization: Launch Vehicles and Programs*, edited by F. Shahrokhi, J. S. Greenberg, and T. Al-Saud, Vol. 126 of *Progress in Astronautics and Aeronautics*, AIAA, Washington, DC ISBN 0-930403-75-4.)

Avery's Career: Solar Energy from the Sea

Avery's career documents the advantages that a solid foundation of knowledge gives to people who undertake energy-related careers. His education began in the local public schools in Colorado and continued through four years at Pomona College. Fellowship grants enabled him to earn a Ph.D. degree in physical chemistry from Harvard University. Somewhere along the way he began a lifelong interest in the phenomenon of combustion, ranging from determining high-temperature combustion rates to designing fuels and hardware for solid-fuel rockets and ramjet-propelled missiles.

Basic knowledge about what energy is and can do opens doors to a wide variety of opportunities to contribute to humanity's well-being. Avery's inventions, for example, include a moving sidewalk that allows people to board at a comfortable speed but then advances to twice walking speed until it slows down for people to exit safely. An aboveground, rapid transit system that mounts four-seater monorail cars on pylons the height of lampposts is another of his inventions. In this system, there is no need to wait for transportation because cars arrive at each station at intervals of a fraction of a minute apart.

Avery took on the problem of insuring adequate sources of available energy when his colleagues conducted a study of alternate energy sources that might alleviate deficiencies. On the basis of this study, it was determined that a possible solution to the nation's diminishing supply of oil and gas exists in the solar energy that warms the top layers of tropical and semitropical oceans. An enormous amount of energy is involved. On an average day, the amount of energy falling on these oceans is 10 times greater than the energy that is obtained worldwide from burning oil throughout an entire year. In fact, more energy falls on these oceans in four or five days than is stored in the earth's entire supply of oil.

Glaser's Career: Solar Energy in Space

Glaser received his early education in Czechoslovakia. In 1939, his parents fled with him to England to escape the Nazi invasion. There he graduated from the Leeds College of Technology with a major interest in mechanical engineering. He then joined the Free Czechoslovakian Army tank corps in England and was transferred to the United States Army unit that was ordered to free Prague from Nazi control. He continued his education at Charles University in Prague but again fled his native country when the Communists took over in 1948. He came to the United States with only the ten dollars permitted refugees who were leaving Prague. Fellowship grants enabled him to earn M.A. and Ph.D. degrees at Columbia University.

Glaser's early interests in energy centered on designing a high-temperature solar furnace at Arthur D. Little. At Columbia, his doctoral research looked into the thermal properties of the particulate material that remained in highly evacuated containers. Early in his career he developed new insulation material that reduced the thickness of refrigerator walls sufficiently to double the inside space. Projects at Arthur D. Little increased his knowledge of the lunar environment and microwave power transmission. These projects included several science experiments that were deployed on the moon during the Apollo program. One ongoing experiment led to the discovery that the moon recedes by two centimeters per year.

Note that energy ties together Glaser's wide array of interests and achievements and that there are good reasons why he became intrigued by the possibility of increasing the availability of solar energy. In this connection, he realized that his research would move along faster if he could have 24 hours of continuous sunlight. This realization became the immediate next step toward developing the space solar power satellite.

Idso's Career: The Greenhouse Effect

Idso's career documents how energy-related issues generate controversy. He stands apart from the popular notion that the continued flood of carbon dioxide into the atmosphere will result in a mind-boggling catastrophe. To him, "humanity's enriching of the air with CO_2 may lead to a great 'greening of the earth,' a biological stimulation of such magnitude that its amplification of the totality of Earth's life processes could only be described as a veritable rebirth of the biosphere."

Idso's education, experience, and personal integrity certainly give him license to defend his point of view and to speak out against those who accuse him of creating obstacles rather than pointing the way toward protecting and improving the quality of life for the world's people. It is not the intent here to enter into this controversy, but it is cited for a definite reason. Idso's defense illustrates how scientific controversy, if properly pursued, can both enrich energy-related careers and improve the chances that society's problems will be solved efficiently.

Idso advises us to examine the reasoning and investigative methods that underlie the entire notion that the continued increase in the CO_2 content of the atmosphere will result in catastrophic warming of the earth. He questions the accuracy of predictions that are based on "complex mathematical constructs designed to simulate our planet's climatic behavior within the innards of a super computer." He sees the global dimensions of the atmosphere, the complexity of nature's processes, and the incomplete understanding of long-term fluctuations in the earth's temperature as providing sufficient reason to question whether the earth's climatic behavior can be accurately reduced to computer programs.

In a second line of reasoning, he advises us to "consider our nearest planetary neighbors, Mars and Venus, whose atmospheres are both within a few percent of being totally composed of pure CO_2. As a result of a number of spacecraft missions to these distant worlds, we know that the greenhouse warming of Venus is approximately 500°C, while that of Mars is only 5 or 6°C." Idso sees these facts, plus the decreasing CO_2 content of the earth's atmosphere over the past four billion years and the increasing luminosity of the sun, as casting further doubt on the accuracy of predictions that are based on mathematical models.

Relationships between clouds and the earth's atmosphere also puzzle Idso.

> Clouds possess both greenhouse and anti-greenhouse properties at one and the same time. That is, they trap and reradiate downwards a certain portion of the thermal radiation emanating from the surface of the Earth that would escape to space and be lost, thereby warming the planet, and they reflect back to space a certain portion of the incoming solar radiation from the sun, thereby cooling the globe.
>
> The net effect of clouds is to cool the surface of the earth. At the same time, cloud cover invariably expands with increasing surface air temperature. Created thereby is an example of the negative feedback whereby new stresses on the environment are dissipated. Another part of this argument involves the type of clouds that occur. In general, clouds that are composed of water droplets last longer than those that are composed of ice crystals with obvious feedback effects. In short, if the earth's temperature rises, there will be more clouds; but increasing the cloud cover will cool the earth.

Advancing the science and technology of agriculture—the endeavor that feeds the world's people—is what Idso's career is all about. To him, nothing is more vital than understanding nature's forces that collect and store the sun's energy. Photosynthesis makes the biosphere possible, and carbon dioxide is its basic raw material—carbon dioxide that is taken primarily out of the air and whose atoms are used as the building blocks for the energy rich substances that form the leaves, stems, roots, fruits, and all other plant tissues and organs.

It follows that all of the carbon that has been locked up in the world's coal, oil, and gas during the billions of years of life's existence on earth, existed at one time in the carbon dioxide content of the atmosphere. In effect, burning these fossil fuels simply returns to the atmosphere the carbon dioxide that made possible the lush biosphere during their formation. With this possibility in mind, Idso asks, "Is this, then, humanity's redeeming grace, the one good deed we have performed (albeit unknowingly) for the other life forms with which we share the planet? Is our returning to the atmosphere of a sizeable fraction of the life-giving carbon locked away in the bowels of the earth so many millenia ago our great redemptive environmental act? I believe that it is, and that it will ultimately prove the salvation of our entire global life support system." ("Carbon Dioxide and Global Change: End of Nature or Rebirth of the Biosphere?," *Rational Readings on Environmental Concerns,* Van Nostrand Reinhold, 1992.

Energy plays such a vital role in economic and sociopolitical affairs that issues involving its utilization and conservation become intermingled among vested interests and ambitions. It is difficult to get at root causes and to sort out proposed solutions to problems. Idso's career illustrates the values that derive from stimulating controversy as a means of bringing knowledge to bear on developing the ways and means for solving energy-related problems.

ENERGY AND THE POPULATION PROBLEM

Many of the earth's energy resources are not renewable. The worldwide supply of coal, gas, and oil, for example, is now in place. These fossil fuels are not being replenished, and the current supply must serve present and future populations. But populations are not finite. A vital part of estimating the adequacy of our energy resources is finding ways to stay up on how many people there are likely to be in our population by the time plans for developing or conserving the earth's energy resources are to be carried out.

Let's play a bit with the problems faced by people who are called upon to make population estimates. Such estimates are based in part on how rapidly people are having children today and how rapidly previous generations reproduced. For example, in the United States today the average life span is about seventy-six years. To make the arithmetic easy, let's estimate the total population to be 250 million and assume that there is the same number of people in each age group. On this basis, about 3.3 million people can be expected to die each year.

To hold the population constant, it follows that there would need to be 3.3 million births each year. Assume that those couples between the ages of 18 and 38 make up the child-bearing group. This age group represents about three-tenths of the total population or 36.5 million couples. For the total population to stay at 250 million, fewer than one in ten of these couples could have children in any one year. Furthermore, assuming that all couples were to have children sooner or later, no couple could have more than two children.

Such a birthrate is not the usual state of affairs. Most human populations are increasing. But current facts suggest that the rate of increase can be slowed down by population control efforts. It is quite difficult, however, to predict accurately how many people there will be in the world in the future. Some scholars have based their predictions on how rapidly the world human population has increased in the past. One estimator believes there were 1,000 people in 1 million B.C.. By 10,000 B.C., he estimates the total world population to have reached 1 million, 275 million by 0 B.C., and by 1970, he says that the world population had reached more than 3 billion.

These estimates suggest that early in human history it took 100,000 years for the world population to double. Near the advent of the Christian Era, the world population doubled each thousand years. It has been doubling in a smaller number of years ever since. Using data collected by the United Nations, world population doubles nowadays in roughly forty years.

If nothing happens to slow this rate of increase, by the time a child who is born today has lived to be forty years old, her or his world will be twice as crowded as it was when the child was born and nearly four times more crowded by the time of old age. In language more meaningful to people who will be working in energy-related careers, by the time today's children reach age 80 they will find four times more people driving their cars into gasoline filling stations, plugging electrical gadgets into outlets, pumping oil or gas into their furnaces, and doing all of the other energy-consuming things that go with living.

People who are looking forward to energy-related jobs and careers may find that their work involves topics ordinarily thought to be the concerns of social scientists and politicians. Suppose, for example, no way is found to head off the diminishing supplies of available energy, and we are forced to cut back on the amount of energy each of us is permitted to consume. Could we change the way we live, the ways we are in the habit of doing things? Would we adjust to using no more energy than people did a hundred years ago? Or five hundred? Or a thousand?

THE ENERGY SUPPLY AND OUR LIFE-STYLES

What would the energy picture look like if you could roll back the calendar 120 years? At that time an average person in the United States consumed about five gallons of petroleum per year, whereas today's average per capita consumption is more than a thousand gallons. Today's average person uses three times as much coal as in 1870; and the average person in 1870 scarcely knew what natural gas was. All in all, the average per capita energy consumption is approximately ten times what it was a hundred years ago.

Look at the way people lived one hundred years ago and the kinds of things they were able to do. Look at cathedrals and castles, paintings and sculptures, highways and bridges, theaters and sports arenas, printing presses and wineries, spinning mills and foundries. All of these achievements were done before the earth's store of coal, oil, and gas had been much more than barely sampled. Coal had been introduced as an energy source as early as 1500, but muscle power and wood provided the major energy supply throughout the world until well into the nineteenth century.

Terribly complex problems face the people whose careers involve the sociological impact that an energy shortage would have on the way we live. It is hard to believe that we would ever allow the stream of civilization to slow down. It is reassuring to realize that hundreds of years ago, even without the massive quantity of energy that characterizes the so-called Age of Fossil Fuels, men and women were able to produce great architecture, musical and artistic masterpieces, functional homes and hospitals, long-lasting bridges and highways, and generally effective solutions to most of their technological problems. But serious problems would surely worsen in any society were its people forced to cut back on their energy consumption. The withdrawal symptoms accompanying a ban on many of our work-saving, time-saving, and standard-of-living-improving gadgets and products could well throw a society into chaos or revolution. What greater challenges are there than those that await people who prepare for careers in the combined fields of energy and sociology!

Many people believe strongly that the challenge of energy shortages will be met successfully. Quoting William S. Sneath of the Union Carbide Corporation, "In my judgment, we can reject the notion that mankind will one day wake up to find a barren planet and an empty cupboard, as the doomsayers seem to fear. I do not challenge their arithmetic, but their conclusions are wrong because they have failed to account for the effects of innovation and imagination. I have worked in industry for almost thirty years, and on countless occasions I have seen the arithmetic of scarcity confounded by these magical ingredients." Those two words, *innovation* and *imagination,* suggest some of the qualities that are needed in those aiming for careers in the energy field today. New ideas, new approaches, and new solutions to problems are much needed.

ENERGY AWARENESS

Preparing for an energy-related career calls for a unique kind of planning. It is not a case of completing a set pattern of courses such as may be required to be a lawyer, a secretary, an engineer, or a chemist. One begins by gaining the knowledge and skills required to become a lawyer, a secretary, or an engineer and then goes on to specialize in energy-related topics. Here is where your own energy awareness comes into play.

It is not enough to be only a good economist, carpenter, or geologist in the usual way we go about preparing for jobs and careers. Those who seek energy-related careers must feel at ease with the language used to describe the magnitude of energy resources, how these resources are discovered and developed, and how the energy they make available can be distributed to wherever it is needed. We need to know something of the scientific laws that describe how energy can be changed from one form to another. Probably most important of all, we need to be aware of our total dependence on adequate supplies of energy for anything and everything we want to do.

This fact was brought home to the author very dramatically by an incident that happened in the public square of a picturesque Bavarian town. The town clock began striking the twelve notes of high noon. Townspeople and tourists paused to look up at the ornate bell tower of the city hall. Carved and painted figures appeared and performed their noontime show. Musicians made music. Cobblers cobbled shoes. Tailors sewed seams. Cabinetmakers wielded hammers and saws. Goosegirls and goatherds tended their flocks. Kings and princesses, mayors and councilmen carried out their governmental duties.

When the clockwork machinery ran down, the show stopped. This was not surprising. Everyone knew that the huge windup toy on the bell tower would run only until it ran down. The pretty scene on the tower happened only because someone wound the mechanism, and the wheels turned and the levers moved only until the stored energy was spent.

Then came the realization that everything the real people were doing in the town square was very much like the action of a giant windup toy. Each person, gadget, or animal that moves or causes anything else to move is equally dependent upon the expenditure of stored energy. Life itself is dependent upon energy transformation. Energy runs the whole world of people, animals, plants, machines, gadgets—even the earth's clouds, winds, and waves.

It isn't always easy to sort out how energy ties together all of humanity's activities. One can see tall, modern high-rise buildings being built in Moscow, Washington, London, or Bangkok. At first glance, each building seems to be the creation of the same energy-hungry bulldozers, cranes, and other massive construction equipment. Closer looks bring out significant differences in energy consumption. In Washington, each bundle of bricks, each window is hoisted to where it is needed by powerful machines. Sometimes the same giant crane is used to lift a steel beam weighing many tons and a handful of rivets weighing only a few ounces. In Moscow, on the other hand, one sees men and women using pulleys to hoist, hand over hand, the bricks and mortar they need to finish a building after the energy-consuming heavy equipment has moved on to other building sites. And in Bangkok, although the facade of a new high-rise office building looks very much like office buildings in Washington, women use hoes and shovels to mix the mortar they need to put the finishing touches on the tile and marblework inside.

High-rise buildings, clockwork machinery, massive construction equipment, winds and clouds, bricks and mortar, giant cranes—a strange jumble of words to try to communicate the idea of energy awareness. Nor is it easy to put into words the idea that the story of civilization is the story of humanity's use or abuse of the earth's energy resources—the happy story of wise use and conservation or the tragic tale of unwise, short-sighted waste. But these are the kinds of ideas that provide the unique challenge and appeal of energy-related careers.

Alec Flamm, an executive officer of the Union Carbide Company, once addressed several thousand engineers, scientists, and technicians. He wanted to emphasize that although technology is very much involved in the solutions to many problems in our society, in the minds of some people, it is equally involved in the causes of problems. He pointed out that, to some people, technology, "is seen as an uncontrolled mechanism that gobbles up huge quantities of our precious resources—energy especially—and leaves in its wake the flotsam and jetsam of a polluted environment."

Flamm used a familiar story to get across his idea of the challenge that faced the people in his audience. It was, "that ancient story of the medieval stone carvers who were asked what they were doing. One said that he was carving a stone; the other replied that he was building a cathedral. Both were correct, of course, but one saw the larger purpose of his efforts."

Flamm closed his speech by telling the engineers, scientists, and technicians that they must take a broader view of how their skills and talents are to be employed in using technology to advance our society. That is, "We must build cathedrals, not carve stones."

To become fully aware of energy's role in our own as well as in the world's affairs can change a meaningless, humdrum job into a truly satisfying career. This is one good reason why potential engineers, lawyers, computer programmers, accountants and dozens of other kinds of workers should keep an eye open for opportunities to gain energy-related employment.

CHAPTER 2

ENERGY: WHAT IT IS AND WHAT IT DOES

Energy is a strange word. We all have a good idea of its meaning, but energy is a very difficult word to define. We know what it means to run out of energy when we have work to do. We know that food and rest will restore our energy. We know we can't move something that takes more energy than we can come up with. We know how comfortable it is to bask in solar energy and how catastrophic the energy that builds up in tornadoes and hurricanes can be.

We know that fuels release the energy needed to warm our homes and to cool or cook our food. We also know that liquid fuels provide the energy for autos and trucks, boats and planes, bulldozers and trains. We know that waiting behind each electric outlet is something that will energize all manner of gadgets. And we know that demolition experts carry enormous amounts of energy in small packages of dynamite or other explosives.

DEFINING ENERGY

It is something else to come up with a definition that really says what energy is and, at the same time, is as precise as a good definition should be. Dictionary definitions confront us with a wide array of terms and phrases: "Imaginative or affective force; vitality; the capacity of acting, operating, or producing an effect; inherent power; vigorousness."

If we continue to seek a dictionary definition of energy we come to this rather awesome definition from *Webster's Third:*

> An entity rated as the most fundamental of all physical concepts and usually regarded as the equivalent of or the capacity for doing work either being associated with material bodies (as a coiled spring or speeding train) or having an existence independent of matter (as light or X rays traversing a vacuum) its physical dimensions being the same as those of work ML^2 divided by T^2 where M is mass, L is length, and T time, usually being expressed in work units (as foot-pounds or ergs) and in any form being endowed with the properties of mass (as inertia, momentum, gravitation) by relativity which assigns to the energy E the equivalent mass m by the equation m equals E divided by c^2 where c is the speed of light.

Although this kind of definition is needed when physicists talk to each other, to demand such a precise definition when we are dealing with energy-related careers could be more frightening than enlightening.

As early as possible each person who is looking forward to an energy-related career should try to put together a definition of energy that will serve her or his needs. Quite often, however, we need to understand the definitions of energy that meet the needs of our fellow workers.

A widely used definition says that energy is the ability to do work. But this definition does little to help us understand how energy enables us to do work, live and grow, stay warm or cool, play or sleep, communicate, or do all of the other things that we do.

Observing Energy's Effects

Energy may be almost totally invisible in many of its forms. Only when energy is changing from one form to another do we see, hear, or feel its effects. Think of an ordinary match, for example. Not until the match is struck and starts to burn do we see and feel the energy it contained.

Energy: What It Is, What It Does 23

We learn about energy by observing its effects on materials. When cold water absorbs energy its temperature rises. When an ice cube melts, energy is being absorbed. Keep adding heat to boiling water and more and more water changes to steam.

A heavy box lifted to a high shelf may look as it did when sitting on the floor. Let the box fall from the shelf, however, and we know that there was an important difference between the box on the shelf and the box on the floor. Our science teachers may have tried to describe this difference in terms of "potential" versus "kinetic" energy, but such a description may cover up rather than reveal the true nature of energy.

We know that food provides the energy we need to live and grow. But there is no way to see in a steak sandwich or a bowl of soup the energy that keeps our bodies warm and enables our muscles to push and pull. We are totally mystified by the energy transformations that allow our eyes, ears, and other sense organs to pick up energy in the form of light and sound waves and then convert this energy into intelligent messages.

Slightly less mysterious is the energy in dry cells, batteries, and electrical circuits. Wires look the same whether or not they are carrying electricity. Let the electricity jump to a finger, however, and we become immediately aware of this form of energy. A new battery can do things a worn-out battery can't do, even though both batteries look very much alike.

X-ray energy will pass through materials that are totally opaque to ordinary light. So will the form of energy that is used to broadcast radio and TV programs. To inadvertently touch a radio antenna and notice an increase in the volume of the music being played may help us realize that the energy being used to carry the program from the broadcasting station passes through our bodies. Because the human body is a good conductor of the form of energy used to transmit radio programs, the radio's antenna picks up a greater amount of the energy coming from the broadcast station when we touch the antenna.

When trying to put together a personal definition of energy, keep in mind that what energy does is the best clue to what energy is. Heat energy, for example, melts solids and changes liquids to gases or vapors. A well-accepted theory, the kinetic molecular theory, explains these effects by saying that heat and the motion of molecules are interrelated. Add heat to a substance and its molecules move faster. Solids warmed above their melting temperatures melt because their molecules are moving too fast to hold a rigid structure. Liquids warmed above their boiling temperatures vaporize because their molecules are moving too fast to stay in touch with each other, as they do in liquids.

Relation to Motion

This brings out another feature that a good definition of energy should include: Energy is closely related to all forms of motion and it matters not whether what is moving is a molecule, a mouse, a sunbeam, or an eighteen-wheel tractor-trailer.

But the energy that leaves a burning match doesn't depend upon how fast the molecules of the match are moving. Somehow our definition of energy must include the role energy plays in holding atoms together in molecules and allowing the atoms in one kind of molecule to be reshuffled into new kinds of molecules. In some cases, energy must be absorbed in order to put together new molecules. In other kinds of reshuffling of atoms, energy is released when fragments of other molecules are rearranged to form new molecules.

Not all of the energy of the burning match is transformed into heat and light. Some of this energy stays in the carbon dioxide, water, and other elements produced by the match's combustion. The usable heat from burning fuels is only a part of the total energy that is involved. From the chemical energy originally present in the fuel we must subtract the heat added to set the fuel burning and the energy needed to put together the molecules of ashes before we have the net energy "profit" from burning the fuel.

Atomic Energy

There is another form in which energy occurs that must be covered by our personal definition of energy. This is the energy that is involved in the arrangement of electrons, protons, neutrons, and other subatomic particles that make up atoms. Split an atom, for example, and energy is involved. Crunch or fuse small atoms together to form a larger atom and again energy is involved—energy in fantastically large amounts when compared with the amounts of energy involved in the motion of molecules or in the reshuffling of atoms from molecule to molecule.

How energy is released when atoms come apart or fuse together can boggle the minds of most people. We know that there is always a loss of mass—that is, the weight of the split or fused atoms is always less than the weight of the original atomic fuel. And this lost mass is converted to energy. These facts were covered in the dictionary definition of energy by the famous Einstein equation, E equals MC^2. This equation enables scientists to calculate the amount of energy that will be released when atomic fusion or fission reactions are used to make energy available.

Many years of study are needed before we can really understand how energy is involved in the makeup of atoms and how this energy is released during atomic reactions. But this does not mean that people who are looking forward to jobs or careers in energy fields can afford to omit this feature from their personal definitions of energy.

In some ways, electrical energy is related to atomic or nuclear energy. Electricity depends upon the flow of electrons. Electrons are the bundles of energy that make up the outer portions of atoms. Energy is involved whenever electrons are pulled away from the positively charged protons in the nuclei of atoms or when electrons are permitted to return to their normal distances from their nuclei.

Again, how electrons absorb energy when they are separated from atomic nuclei is difficult to understand. So is how they give up energy when they return to "ground state." This phenomenon becomes another part of our definition of energy.

Equally difficult to understand is how energy is related to gravitational forces. For sure, we know that unsupported objects go crashing to the floor. To hear the thunderous roar of waterfalls and to see falling water drive electrical generators convinces us that there is a close relationship between gravitational forces and energy. To realize that winds are large masses of air moving in response to gravitational forces provides further evidence of the link between energy and gravitational forces. So does the ebb and flow of tides and the overwhelming power of wind-driven waves.

Do you see the many gaps that are left by the definition that energy is the ability to do work? To fill these gaps, we need to build a definition that covers the many forms of energy and the many things energy does when it is changed from one form to another. Whatever energy is, it is closely related to the motion of objects, the position of objects, and the makeup of atoms and molecules. And it could well be that a good personal definition merely emphasizes the idea that energy is what energy does.

ENERGY MEASURING UNITS AND HOW TO EXCHANGE THEM

Much confusion comes from using several different kinds of units to measure the amount of energy in an energy resource or to pay our energy bills.

We buy coal by the ton and gasoline by the gallon, and the energy these fuels contain is often expressed as British thermal units (Btus) per ton or per gallon. Our gas bills may be figured as so many units (Btus) per thousand cubic feet or as so many therms (100,000 Btus). Electric bills are calculated as so many kilowatt hours. We might measure the amount of energy needed to climb a set of stairs in foot-pounds. The energy available from the engine in an automobile can be measured in units of horsepower. The amount of solar energy falling on the earth is measured in calories per square meter or Btus per square foot, although people can add to the confusion by intermingling these units and talking about calories per square foot or Btus per square meter.

To be at ease using a variety of units to measure energy is something to be accomplished early in preparing for energy-related jobs and careers. The problems that must be solved are similar to converting money when we travel from one country to another. If we are in London, for example, and need to buy a new pair of shoes, we know that the cost is the same no matter whether we pay for them in dollars or pounds sterling. We also know that we have only so much money and it doesn't matter whether the amount is expressed in dollars or in pounds.

Much of the confusion that comes from using different kinds of "coins" to pay for the energy we use can be avoided if we keep in mind that the amount of energy involved in any specific energy transaction will be the same no matter what kind of coin we use.

For example, it could become possible to think of the 100 food calories available in a slice of bread as being roughly 400 Btus, 300 thousand foot-pounds, 0.12 kilowatt hours, 420 thousand joules, or 4.2 trillion ergs. The actual amount of energy stays the same and only the value of the coin being used changes. More will be said about exchanging one energy measuring unit for another in later chapters. For now, be assured that the problems seem larger than they really are.

HOW SCIENTISTS DESCRIBE ENERGY

There is one more hurdle to be jumped on the way to getting a feel for what energy is. We need to appreciate the efforts scientists have spent trying to describe the properties or behavior of energy. The scientific laws that summarize the properties and behavior of energy are delightfully simple and, at the same time, overwhelmingly sophisticated.

We are assuming here that simple interpretations of these laws will serve to help people look forward to energy-related careers. Sophisticated interpretations are postponed until training programs may call for more understanding of subtle differences.

Probably the most characteristic of the properties of energy is that it simply won't stay put unless it is "locked up" by some kind of interaction with matter. Put a warm object in a cool environment and heat will leave the warm object. Each time we use a vacuum bottle to keep beverages hot or cold we realize that special efforts are needed to keep heat energy from moving to cooler environments. Light a match and the match burns and keeps on burning until it is burned up. Start any chemical reaction that allows energy to leave the reactants and the reaction continues until at least one of the reactants is totally consumed. Ignite the explosive mixture in the cylinders of an automobile engine and the energy in the fuel is sure to be transformed into heat and the motion of the pistons until the fuel or the oxygen provided by the carburetor is totally consumed.

Another property is revealed each time a quantity of energy is changed from one form to another. Each energy transformation allows a fraction of the energy to escape, never again to be recycled or regained. Supposedly, far into the future, all of the earth's energy will have escaped or reached that mysterious state known as entropy.

The notion of entropy stretches our minds and imaginations almost to the breaking point. But this notion has extremely important implications for people who are interested in energy-related jobs and careers. This is true for people who work directly in producing and distributing energy as well as for people whose jobs are in the economic, social, or political aspects of the energy industry.

The concept of entropy forces us to realize that the earth's energy resources are dwindling relentlessly. It is true that the sun replenishes our energy supply, but the supply of energy in the form of coal, oil, gas, and uranium is another matter. It could be that far into the future the energy available in these fuels will have been released. If this happens, only solar energy, the energy in gravitational forces, and new sources of atomic energy (fusion, perhaps) will be available to keep running the giant windup toy that is the earth.

Our dependence upon the flow of the sun's energy through our environment shows most clearly in the world of green plants. By that wonderful process, photosynthesis, the sun's energy is trapped, so to speak, in the complex molecules of sugars and starches that are put together by green plants. When the chemistry of life's processes takes these complex molecules apart in the bodies of plants and animals, energy is released. Before this energy can continue on its path toward entropy, it provides the energy that is essential for life.

ENERGY AND THE ENVIRONMENT

To get a feel for what energy is, we must think about how energy is related to the pollution of our environment. This relationship is closely tied to the relentless restlessness of energy. A second factor is the tendency for things in nature to scatter about. This may appear to be a strange observation, but in reality it describes a fact of nature that we see quite often.

Jiggle a castle built from toy blocks and the blocks fall into a jumbled mess. Spill a new deck of playing cards and, although the cards were sorted by suit and denomination, it is very, very unlikely they will remain that way after being spilled. Contrast the building of a high-rise brick building with what happens during the demolition of the same building. Workmen go to considerable effort to put each brick into proper order, but allow the building to tumble down and all we have is a pile of bricks.

The point here is that energy can be trapped by using it to put things in order. But the restlessness of energy shows through the moment anything happens that allows the ordered system to become disordered and, thus, allows the energy to escape and continue on toward entropy.

Now think of a lump of coal or a drop of gasoline. Here we find atoms and molecules arranged in highly ordered systems. Allow the coal to burn or the gasoline to explode and these same atoms and molecules can be spread far and wide while the energy originally present in the coal or gasoline is being changed into heat.

Metals provide another example of order versus disorder. Metallic ores are often in the form of precisely put together crystals. In some cases—copper, for example—chunks of the pure metal occur in the form of highly ordered crystals. Suppose some of this copper is extracted from the ore and then dissolved in acid to make a compound to be used to clean aquariums. Sooner or later this copper could well be scattered throughout the world's oceans. Much energy would be needed to retrieve the copper and put it back into the form of the highly ordered crystals of valuable copper ore.

This is really what environmental pollution is all about. The oft-repeated law of conservation of matter and energy assures us that none of the earth's mineral resources can ever be destroyed (without converting them to energy). But this law says nothing about the relentless escape of energy to entropy, a condition which makes energy useless to us. Nor does the law of conservation say anything about the energy that must be spent to put scattered materials back into useful form.

Pollution is scattering and scattering is easy. Cleaning up is putting things back in order and this takes energy. There is no way for people who are looking forward to jobs or careers in the energy field to avoid becoming equally involved in the problems that grow out of the increasing pollution of our environment.

In nature, enormous amounts of solar energy have been packed into the ordered atoms and molecules of our fossil fuels. For the most part, this happened millions of years ago, and it was pretty much a one-time event. The energy that is stored in these fuels is conveniently available. But this energy moves one step closer to never again being recycled when the fuels are burned and the highly ordered atoms in their molecules are scattered far and wide in exhaust gases.

There have been and always will be great challenges awaiting people who are willing to prepare for careers in the energy field. Much of the complexity of these challenges lies in the very nature of energy itself. It would be misleading to gloss over the amount of study needed to really understand the true nature of energy as a scientific topic. Actually, success in many energy-related jobs depends simply upon constantly staying alert to what energy does even though there is a great challenge to understand what energy really is.

CHAPTER 3

CAREERS IN THE OIL INDUSTRY

From many points of view, petroleum is the world's most important commodity—economically, politically, and sociologically. Its greatest use is as an energy source, but it is also the preferred raw material for creating thousands of substances such as plastics, textile fibers, fertilizers, drugs, insecticides, and detergents. An advertisement by a company that specializes in petrochemicals carried the headline: "Besides gas for your car and heat for your home, can you name a few other things oil is used for?" The ad went on to help the reader by listing 497 other uses from baby bottles to windshield wipers, guitar strings to panty hose.

Petroleum illustrates nicely how the energy industries employ people with widely different career interests. One major oil company, for example, lists job opportunities for people who are trained in accounting, business administration, chemistry, computer science, economics, engineering (all branches), finance, geology, geophysics, liberal arts, marketing, mathematics (statistics), law (petroleum land management), and physics.

THE NATURE OF THIS ENERGY RESOURCE

Each job in the oil industry can be traced to the properties that make petroleum useful either as an energy resource or as a feedstock for making marketable materials. On this basis, each potential employee can gain perspective from knowing something about what oil is, how it was formed, and where oil deposits fit into the scheme of the earth's geology.

The consensus of opinion is that petroleum was formed millions of years ago by the partial decomposition of plant and animal debris. Efforts to trace the origin of petroleum begin with the belief that large areas of the earth's crust have undergone striking changes during the millions of years since the planet was formed. Alternate sinkings and rising of crustal blocks have been accompanied by extreme changes in climate together with repeated relocation of the oceans. Large areas that at one time were covered with dense growths of vegetation were gradually submerged until they were completely covered by an invading ocean. While these crustal blocks were submerged they collected thick layers of oceanic sediments.

The rock strata we see exposed along road cuts or canyon walls differ in color and texture because environmental conditions differed during the submergence or emergence of that part of the world. Supposedly, in the distant past, there were areas where heavy growths of semitropical swamp plants accumulated while the areas were submerging. When these areas became totally submerged, the not yet entirely decayed plant and animal debris became covered over by thick layers of ocean sediments. In effect, a very fortunate set of circumstances created massive chemical factories where onetime organic material would be slowly taken apart and new substances created.

An extremely important feature of green plant growth is that the sun's energy is stored, so to speak, in the leaves, stems, and other tissues of the plants. While petroleum was being formed, a fortunate set of circumstances caused much of this energy to be trapped in the buried debris. Ordinarily, when organic materials decay, the large molecules of their formerly living tissues come apart, new smaller molecules are formed, and much of the energy originally present escapes. But without enough air to provide oxygen, the usual decay products could not be formed. Eventually, enormous amounts of buried debris were converted to hydrocarbon compounds—compounds we now know as petroleum and natural gas. The formation of coal differs only in that more of the hydrogen was removed and the end product was left with a high carbon content.

The newly formed hydrocarbons, both gases and liquids, diffused throughout the porous layers of sediment that were being compacted to form rocks. Some kinds of rocks allowed the fluids to flow easily whereas others were almost totally impervious to the oil or gas and served to trap the hydrocarbons. The gas- and oil-soaked rock strata become the pools which, upon being discovered and tapped, become the oil and gas fields of today.

There were probably as many high and low places in those prehistoric landscapes as there are in similar swampy areas of the world today. Small lakes probably occupied the low spots and a dry land biota took the place of swampy growths on the higher areas. This may be why different portions of an oil field may produce different kinds and amounts of oil and why dry holes continue to surprise and frustrate petroleum geologists. Furthermore, because different kinds of rock strata were formed above the gas and oil pools, much of the oil and gas may have diffused far from where it was formed and may even have escaped to the surface and evaporated. The expertise of locating oil and gas deposits hinges on being able to reconstruct in one's mind how these ancient landscapes were millions of years ago.

The geological stresses that cause the earth's crustal blocks to sink or emerge also cause enormous horizontal displacements of rock strata. Road cuts through mountain areas often reveal "layer cake" rock strata tilted into all manner of ups and downs. In areas where earthquakes occur, fault lines show where one crustal block has been forced to ride over adjacent crustal blocks. To see these badly rearranged rock strata and to realize that gas and oil exist only where underground strata create traps for them helps us to appreciate the skill and knowledge that enable prospectors to predict where oil and gas fields might be discovered.

When a well is drilled into layers of rock that have trapped gas and oil, the pressure these substances are under causes them to flow toward the surface. In the early days of the oil industry, new wells sometimes struck formations where the oil and gas were under so much pressure that large volumes gushed upward forming huge fountains that rose hundreds of feet into the air. Today's technology, however, provides drillers with ways to control the flow of gas and oil from newly opened wells.

The repeated sinkings and risings of the earth's crustal blocks, together with tilting and sliding of these blocks, have caused rock strata that are deep underground in some parts of the world to be exposed in other locations. When these exposed strata contained oil or gas, much of it diffused and evaporated. Most of these hydrocarbons were lost to the atmosphere many years ago, but in some parts of the world the continued seepage produced the first natural oil and gas wells.

Actually, these brief paragraphs may be more misleading than enlightening. Much of the action we are describing took place more than 300 million years ago. The plants and animals we mentioned flourished in an environment that existed only at that time. The sinkings and rising of the earth's crustal blocks created the enormous mountain ranges that we know today as the Appalachians, Rockies, the Alps, and the Andes. What was once a single landmass, which scientists call Gondwana, has since been broken up into segments that have become today's continents—continents that continue to drift as though some mighty hand insists on moving the Americas west from Africa, Asia southward, and Australia toward the north.

Terribly complex circumstances determine where oil is to be found today, and in what quantities. Not only has its original location been displaced worldwide, it exists today only where a favorable set of circumstances have prevented its exposure to the air and subsequent evaporation. Needless to say, the ability to determine where these sets of circumstances came together separates the successful petroleum geologists from the unsuccessful.

In some respects, it is easier to determine what parts of the world are unlikely to contain oil reservoirs. For sure, those areas of old Gondwana that were limited to the far northern or southern latitudes, areas that would not allow a biosphere to flourish, offer poor prospects. As seen by C. D. Masters and his colleagues at the U.S.

Geological Survey, "The future of oil production in the Western Hemisphere is anchored in Venezuela. There, in middle Cretaceous time, superior source sediments, known as La Luna, were deposited under restricted geographic conditions that were unique, at that approximate time, to that general area and only a couple of other areas in North Africa and in the Middle East."

Masters also calls attention to the "enormous quantities of unconventional crude oil" that are potentially available from the Athabasca and related tar sands of Canada. Differences in Middle Eastern, Venezuelan, and Canadian oil resources point toward the effectiveness of the trap rocks that permit or retard the escape of the more volatile components of petroleum. The extra-heavy oils of Venezuela indicate that they have been less well trapped than the light Middle Eastern oil. The oil shale and tar sands of Canada have retained the least volatile components.

A more complete discussion of these highly important features of careers in the oil industry appears in "World Oil and Gas Resources—Future Production Realities," by C. D. Masters, D. H. Root, and E. D. Attanasi, and published in the *Annual Review of Energy*, 15:23–51, 1990.

HISTORY OF THE PETROLEUM INDUSTRY

Early Uses

The earliest uses of petroleum products are suggested by archaeological evidence of the prehistoric use of asphalt or pitch to caulk ships and boats. This thick, gooey residue from the evaporation of petroleum provided an easily obtained, water-resistant material to plug leaks.

Archaeologists also tell us that the axles of the pharaohs' chariots in ancient Egypt were probably lubricated by a crudely refined petroleum product. The ancient Arabs knew how to make medicines from petroleum and how to extract products from petroleum to use in lighting their homes and in cleaning clothes. "Eternal fires" and "burning springs" (which we now know were caused by gas or oil seeping to the surface) played interesting roles in primitive religious rites.

By the mid-1800s, a few people were collecting petroleum from seepages or shallow wells and using it to make liquid fuels, as well as the infamous "snake oil" sold by carnival pitchmen. In general, however, humanity lived in the presence of oil and gas from the beginnings of time until about 1840 without realizing how great an energy resource petroleum could be.

Industry Beginnings: Petroleum in Kerosene

According to historians, it was a shortage of whale oil in the United States in about 1840 that led people to seek a substitute fuel for lamps. Although as many as 700 whaling ships tried to meet the demand for whale oil for the American market, the supply dwindled until the price became prohibitive. It is interesting in this connection to note that more than once in our history have whales become an endangered species. To meet this first energy crisis, other oils were substituted. People tried oils from turpentine and from animal fats. A type of kerosene was made from soft coal. None of these substitutes, however, was a truly satisfactory replacement for whale oil.

Near Titusville, in northwestern Pennsylvania, oil seepage could be skimmed from the surface of a small stream known as Oil Creek. In 1859 Edwin K. Drake and W. A. Smith drilled a hole near this creek and struck oil at a depth of about 70 feet. Drake and his partner had been hired to drill the well by a group of businessmen who knew that a good type of lamp oil could be distilled from oil. These men foresaw that if there was an abundant supply of oil, the new lamp oil could be produced at a cost much lower than that of whale oil.

Within months after the first oil well was discovered, hundreds of wells were being drilled in the surrounding area. What started out as a small trickle of oil rapidly became an enormous stream. Getting the oil to market became more of a problem than getting it from the underground strata where it was stored. A good market developed for barrels to transport the oil and for large wooden tanks to store the oil until it could be distilled and refined.

The first refinery began operating in the Titusville area in 1861. It produced lamp oil or kerosene and little else. There was some market for lubricating oils and greases, but the heavy demand for the new lamp oil that was almost odorless and smokeless kept the refineries in business. Within a few years railroad cars were fitted with wooden tanks to haul petroleum, and by 1879 more than 100 miles of pipeline had been built to carry the oil that was pumped over the Allegheny mountains to Williamsport, Pennsylvania.

Soon kerosene oil lamps and cooking and heating stoves became the wave of the future in homes and business places throughout America. Candles and whale oil lamps were rapidly replaced by more efficient kerosene lamps. Wood- or coal-fired cooking stoves lost out to the more convenient oil stoves. People were slower to give up their coal-fired furnaces and wood-burning fireplaces, but oil-fired space heaters gradually took over.

The early years of the oil industry hold valuable lessons for people who are thinking about career choices. The oil industry has always been and always will be based on discovering rock strata which contain oil stored under such conditions that it can be extracted and marketed profitably. And profitability depends upon the cost of the extraction of the crude petroleum, refining costs, distribution and marketing costs, and the price obtained for petroleum products.

Until a market for petroleum developed, this energy resource had little or no value and there were no job or career opportunities identified with it. The availability of petroleum as an energy resource, in turn, generated new markets and thereby created new job and career opportunities.

Petroleum in Gasoline

These facts were illustrated spectacularly by a series of events that began when several inventive people began tinkering in obscure sheds with building "horseless carriages." In 1892 Frank and Charles Duryea built the first American gasoline-powered automobile. Other horseless carriage inventors were using steam or electricity to replace the "hay burners" that provided the energy needed to keep the carriages of the day rolling along.

It took only a few years for gasoline-powered engines not only to send thousands of horses out to pasture but also to beat out the competition from steam- and electric-powered automobiles.

Henry Ford and his mass production of the Model T played a major role in the move toward gasoline-powered automobiles. In 1910 there were fewer than a half million motor vehicles in the United States. Within ten years there were 9 million. During this same ten-year period, a developing oil industry had installed gasoline pumps at nearly every crossroads grocery store or village hardware. People still took their "oil cans" with them when they needed "lamp oil," but storekeepers soon became "filling station operators" as the demand for gasoline grew much faster than the demand for kerosene.

Recalling the early days of the automobile's impact on the oil industry points out additional valuable lessons involving career choices and decisions. When kerosene was the best-selling product of the oil refining industry, gasoline was an almost useless by-product that actually created waste disposal problems. Increasing popularity of automobiles, however, reversed the situation. Soon the refineries could not meet the demand for gasoline, while kerosene became the surplus by-product. Only so much gasoline could be produced from each barrel of crude oil that went through the refinery. Increasing the production of gasoline increased the production of kerosene and other petroleum products as well.

This state of affairs created challenges that were especially inviting to chemists. Chemists knew that petroleum was a mixture of hydrocarbon molecules and that the chief difference between kerosene and gasoline was the number of hydrogen and carbon atoms in their molecules. Gasoline is a mixture of hydrocarbons whose molecules contain approximately eight carbon atoms and eighteen hydrogen atoms. Kerosene and fuel oil consist of molecules with longer chains of carbon atoms and their attached hydrogen atoms. Hydrocarbons with fewer than five or so carbon atoms per molecule are too volatile to use in gasoline.

In 1913 two young chemists, William M. Burton and Robert E. Humphreys, found that when longer chain hydrocarbon molecules are trapped in a high pressure container at a high temperature, the long chain molecules are "cracked" into the shorter chain molecules that are suitable for gasoline. By adopting this cracking process, refineries were able to double the yield of gasoline from crude oil within a few years. It is projects such as this that have created a close relationship between chemistry and many facets of the energy field.

Oil Industry Expansion

The more recent chapters in the history of the oil industry need less detailed explanation to remind us of the rapid expansion of the industry. The good market for gasoline and other petroleum products spurred people toward prospecting for new fields, drilling new wells, and sending petroleum products to market in ever increasing quantities. Improved cracking and other refinery practices continued to increase the yield and quality of the gasoline obtained from crude oil. By remodeling the hydrocarbon molecules and by adding tetraethyl lead or other additives, the tendency of poor quality gasoline to cause engines to knock was practically eliminated.

Automotive engineers have been adept at redesigning automobile carburetors and engines to adjust to the new fuels produced by the refineries. Recalling the automobile models that persisted in "running" after the ignition was turned off illustrates the need for close cooperation between petroleum chemists and automobile design engineers. Petroleum chemists had to describe accurately the burning properties of the fuels their companies produced. Automobile design engineers needed to describe exactly the conditions which develop in the cylinders of an engine when the spark plugs ignite the explosive mixture of vaporized fuel and air.

Wartime Development

Social historians often discuss the impact of the two World Wars on the development of the petroleum industry. The stimulated demand for petroleum products during World War I created production facilities which, when the war was over, enabled people to solve many agricultural, industrial, and transportation problems. New highways surfaced with petroleum asphalt enabled farm people to "get out of the mud." Gasoline-powered tractors replaced horses and mules. Diesel-powered trucks delivered materials directly to individual farms and factories. No longer did people need to pick up freight or parcel post at the local railroad depot or post office.

During World War II, even greater demands were made on the petroleum and petrochemical industries. Not only were existing facilities greatly expanded, but entirely new industries were developed. Huge quantities of high-octane aviation fuel, specialized lubricants, and petrochemicals had to be produced. Entire factories were devoted to producing such products as butadiene for synthetic rubber or toluene for TNT.

Following World War II, peacetime demands for petroleum products increased phenomenally. It was as though a pent-up urge to go places and do things had been unleashed among America's people. Americans took to the highways and airways in droves. In the five years after World War II, automobile registrations increased by 50 percent. So did the number of farm tractors. Uses of oil in addition to transportation also increased. The number of oil-fired home furnaces doubled during the five-year period. Consumption of petroleum products jumped from 1.8 billion barrels in 1946 to 2.4 billion barrels in 1950, and to 6.4 billion barrels by 1976. By 1978, annual consumption of petroleum in the United States reached 7.1 billion barrels.

In the late 1970s, however, the abrupt increases in the price of oil, coupled with problems that developed in the distribution of imported oil, caused the amount of oil being consumed to decrease. In fact, by 1990, consumption was down to 5.3 billion barrels per year.

Abrupt fluctuations in the available supply and market price of petroleum products, particularly motor fuels and heating oil, create what are probably short-term changes in the number of job opportunities in the oil industry. The news media may overplay the impact of "critical shortages" versus "oil gluts." In general, even after admitting that there are close relationships between the number of job opportunities and the immediate supply and demand for petroleum products, it is safe to say that employment in this industry will continue to be reasonably stable.

Petroleum consumption data during the late 1970s and early '80s reveal a very significant fact: the American life-style is not locked into automatically consuming more and more energy. People will respond to efforts to conserve our energy resources, especially when such efforts are accompanied by rising energy costs.

KINDS AND NUMBERS OF JOB OPPORTUNITIES

Approximately 1.5 million people are employed in the petroleum industry. Some of these people locate and drill wells from which crude oil is obtained. Others transport the oil to refineries where it is converted into gasoline, diesel fuel, heating oil, and other petroleum products. In turn, marketing and delivery of these finished products provide many jobs. Finally, a large number of research scientists, lawyers, and public relations experts provide backup services for those who produce, refine, and deliver petroleum products.

Perhaps the best way to sample the wide array of petroleum-related jobs is to adapt bits of conversations among people who fill these jobs at a typical large oil company.

Geologists and Technical Assistants in the Earth Sciences.
"Courses in chemistry, physics, geophysics, and biology, as well as geology help interpret the deep echo-soundings taken from subsurface rock. Then I have to use my imagination to picture what the subsurface looks like and draw three-dimensional maps that show how the rock formations bend and turn in all directions—up and down. And to the right and left. Certain patterns of rock layers may indicate the presence of oil or other valuable resources.

"We make weekly well charts to show how drilling is proceeding. Or we may do some month-to-month studies on the rise and fall of underground water levels and their relation to nearby drilling or testing. On some of these jobs we use a computer, so it's important that I know something about computer operations and what help we can expect from computer services."

Drilling Representatives and Oil Field Operators. "The drilling rep is the company's representative at a drill site wherever it is—on land or sea. You order all of the equipment and services (casing, cement, electric logs, etc.) and check to see that everything gets there on time and the services are performed correctly. You have to make daily reports about drilling progress to your home office and to state and U.S. governmental agencies.

"When the wells have been drilled in an oil field and the mighty pumping units are seesawing away, the field operator patrols constantly on the lookout for pumping unit failures, line leaks, oil spills, and potentially dangerous situations."

Refinery Operators and Refinery Craftsworkers. "The refinery operator is responsible for millions of dollars worth of equipment that produces lubricating oils, gasoline, diesel fuel, jet fuel, or one of the many materials derived from petroleum. Computers make the operator's job more efficient by being connected to instruments that monitor the processes of the refinery. I get information from the computer, analyze it, and then decide what adjustments have to be made.

"The work is demanding and it may be hazardous. Repair sites may be high off the ground. There are explosive gases and high-voltage electricity, and areas where strong acids are used. This is no place to be lazy or careless, but if you follow the safety procedures, you won't get hurt."

Mechanical Engineer. "Mechanical engineers and technicians work closely as a team to help the equipment operators keep the machines running as efficiently as possible. The refinery is a huge, active place with big machinery, tanks, pipes, valves, furnaces, and so on. So we spend some of our time prowling around thinking of new ways to make things safer and more efficient.

"Engineering appeals to people who like to know what makes things tick, whether it's a bicycle gear or a jet engine. Schooling provides the general education and knowledge of engineering theory that is needed to design and improve the basic machinery for practically all industries. Because they have a broad knowledge of technical theory, mechanical engineers understand the 'big picture.' But the technicians have the first-hand, nuts-and-bolts experience of running and repairing their own equipment."

Marketing Representatives. "Marketing reps are responsible for sales, service, collection, and complaints. If you are going to be successful, you've got to like people, be willing to learn, and be able to work on your own, to plan your day, and make your own decisions."

Customer Representatives. "The company computer system is an efficient tool for customer representatives. Customers, who may call from any of the 50 states, are usually surprised that all information about their accounts is at our fingertips and that we can answer their questions without delay. We can update their account, change a customer's billing address, or order a new credit card if one has been lost or stolen. We just type in the correct code—a combination of letters and numbers.

"At first it is hard to deal with some customers, but we learn not to take their dissatisfactions personally. We have to be tactful and choose our words carefully. That means having a good grasp of the English language—both verbal and written. It helps to be organized and to think carefully about each situation. It's a lot easier to be patient with the customers when you're sure about what you are doing."

Service Station Dealers. "Operating a service station really depends on working with people successfully. Sometimes I have to be firm, like making sure that all of my employees start their shifts on time. On the other hand, they may ask advice about their own lives, and then I have to listen carefully and be understanding.

"Gasoline is the main product a service station sells but we also do a lot of mechanical work—minor engine repairs, tuneups, brake work, lubrication, tire changing, and so on. I inspect the work to make sure it's all correct and then go over the bill with the customer to make sure he understands it and is satisfied.

"A successful dealer trains his helpers to be alert for opportunities to sell fan belts, tires, radiator hoses, batteries, and similar car maintenance items. And I must keep an adequate inventory of these items on hand just as I must make sure that the gasoline storage tanks are never pumped dry.

"What with doing the bookkeeping and other paperwork, for the first year I worked harder than I ever did in my life. But I got more satisfaction from succeeding than I had ever felt before."

Chemical Engineers. "In school the chemical engineer-to-be studies chemistry in a lab using a bunsen burner, test tubes, and small amounts of chemicals. Then, in courses such as reactor design and unit operations, he or she learns to translate this small-scale chemistry into what goes on in a chemical plant using gigantic furnaces, tanks, reactors, pipe systems, and huge amounts of chemicals. The translation from small to big is called 'process design.' I've used the principles of process design in many assignments—from designing a plan to remove the high sulfur content of Arabian oil to creating a plant for cleaning up industrial waste water.

"A different approach was used to calculate the costs and the practicality of alternate forms of energy. This called for looking into such energy sources as raising special crops to be burned for fuel, geothermal power, and extracting oil from coal, tar sand, and oil shale.

"We all agree that working in a refinery requires being able to think fast on your feet and be good at working with others. As a troubleshooter, your job is to make sure the complex chemical processes involved in refining crude oil proceed smoothly and safely."

Research Chemists. "Research chemists are hired to create new ideas and to solve problems that are directed toward developing useful, practical materials and processes. Materials such as fibers and fuel. Processes such as making better gasoline. The company pays us to 'think' chemistry. So we do. We think, we read, and we talk to other research chemists about our projects. We also give directions to the laboratory technicians and talk to them about the progress of our experiments.

And then we write reports—lab reports, progress reports, project reports.

"Part of my training as a research chemist was learning to have patience—patience to get through the days when nothing seems to work the way you expect. We have to learn to accept some failure, because a lot of our experiments and ideas just don't work!

"What type of person is drawn to chemistry as a profession? Someone who has an endless curiosity most of all. Someone who is always asking, 'How does that work?' and 'What made that happen?' It really is exciting to run experiments—not only in the lab, but also in your head. You invent models of the problem you are trying to solve. And the best part is, you know you have a good chance of finding the answer."

Computer Operators and Computer Programmers. "The computer is an expensive, delicate, electronic network of machines that don't even like changes in the weather. Although other people in the company communicate with the computer, we are the only ones who actually see it and handle it. It is our job to maintain its delicate environment and keep the computer humming along without interruptions.

"The computer is a machine that follows instructions written in a special language by computer programmers. They write instructions—programs—about the company's business, telling the computer how to do tasks more quickly and accurately than could be done by hand. For example, computers type our paychecks and manuals, address mail for the company, keep track of our oil tankers—they even run the elevators in the buildings! In most cases, the programmer develops a program on a machine that looks like a typewriter with a television attached. The computer records the typed instructions on a disk similar to a phonograph record, which can be stored until it is needed.

"The programmer's job is to learn about a specific business problem and decide how the computer can solve it. Then we give the computer detailed instructions to follow to complete the task. If you are thinking about becoming a computer programmer, it is good to have a logical, inquiring mind. In programming, it's a long slow process finding out if you are a winner, but it's a triumph when you are. If you are to advance in the field, good communication skills are very important. You need to be able to speak and write clearly, and understand the English language."

Information Technicians. "Some information technicians receive data from the scientists, organize them, and put them into computer programs to produce maps or graphs. Others receive various kinds of documents from the scientists, label and classify each document, and assign it a special number. After they record the number, title, and location of the document in a computer, they file the document in a storage room.

"Input means getting it into the computer and retrieval is finding it once it is there. The information we work with includes correspondence, legal agreements, geological reports, area files, and well files. A well file is a record of everything that has been done with a certain well even before it's drilled. I have to understand what's happening out there in the field when they're collecting raw data. That's why I need a background in the earth sciences, too—including geophysics, geochemistry, and paleontology. If I didn't, I wouldn't know what the scientists were talking about when they came in with a request."

Public Affairs Writers. "We deal with the current affairs that concern the public, and 'public' means everyone—people living and working in communities all across the nation. We try to see to it that the public's perception of the company is an accurate one. Communication is our business. Every time an executive makes a speech, a public affairs writer has researched the subject in depth and done a number of drafts.

"We know the rules of punctuation, grammar, and spelling, but we also know how to use these basic skills to produce clear and interesting prose. To get a job as a public affairs writer, you need proof that you can write well. I started working for a newspaper when I was 14 years old. And I continued through high school and college, working and writing for newspapers. Just write, write, write for anything you can. Try to get your stories or articles published. When they do get published, save a clean copy of everything and put it in a folder. Then when you interview for a job, an employer can see what you have written.''

Trends in energy-related industries tend to follow whatever is happening in the overall economic situation. Job opportunities in the oil industry increased rapidly, for example, in response to the oil shortage following the embargo on imported oil during the 1970s. Similarly, employment opportunities in the coal industry responded to the higher cost of oil and gas resulting from the embargo.

EDUCATION AND TRAINING

One way to get started in the petroleum industry is to apply for whatever entry-level jobs are available in one's community. Worthwhile things can be learned and valuable contacts established in any kind of petroleum-related job. Summer jobs also provide opportunities to become known by the people who make petroleum products available to the public. Helpers on exploration and drilling crews obtain valid tastes of what it takes and what it means to pursue those careers.

Climbing the ladder toward increasingly responsible positions requires both experience and specialized training. Experience is the best way to learn how to fit into an ongoing organization and to share the responsibilities that enable teams to work under sometimes hazardous or isolated conditions.

Specialized training begins with the science and mathematics courses usually offered in high schools. In today's world, however, it is becoming equally important to do well in computer technology and in the courses that enable people to follow directions and to write and speak clearly and grammatically.

At the college level, major emphasis during the first two years tends to be on becoming proficient in basic science and mathematics. At Marietta College, in Marietta, Ohio, for example, the first year's program includes "Introduction to Petroleum Industry" together with courses in math, geology, chemistry, physics, English, and speech. The remaining years are devoted primarily to petroleum courses. During the senior year, however, all students are required to take advanced English and advanced speech courses.

Specialized courses cover the things people need to know to qualify for jobs at nearly all levels of the petroleum industry. Course descriptions include such phrases as "fundamental petrophysical properties of reservoir rocks;" "distribution and movement of oil, water, and gas in porous media;" and "calculation of hydrocarbons in place by volumetric methods."

Emphasis on engineering shows through such phrases as "design and optimization of rotary drilling equipment;" "gas well completion techniques;" and "reservoir engineering aspects of waterflooding, gas injection, polymer flooding, micellar solution flooding, CO_2 flooding, steam stimulation, steam flooding, and insitucombustion."

Looking ahead to becoming fully certified Petroleum Engineers, all seniors are required to take (but not necessarily pass) the examination for becoming a registered professional engineer.

Many of the basic science and engineering courses include laboratory work that involves the use of computers. Topics studied and presented, both orally and written, during a senior seminar encourage independent study and research as well as cultivation of written and oral skills.

Universities and technical institutes that are located near major oil-producing areas often offer courses especially designed to prepare people for jobs in the industry. At the University of Texas, for example, students can take courses in drilling operations; well planning, including drilling fluids, hydraulics, and drill-stem components; cementing practices; safety regulations; hoisting systems; human relations; and similar on-the-job skills.

There are also courses that provide a general introduction to onshore and offshore operation procedures. At the basic skill level, there are courses on the maintenance of electronic and electrical systems, automation technology, and the properties of the particular materials involved in the production and marketing of petroleum.

Courses in pipeline technology provide a basic understanding of pipeline construction, maintenance, and management. Other topics that are covered in these courses include arranging contracts for the right to build pipelines through private property, the installation of pumps and meters, and the design and operation of terminal facilities.

Many people believe it is wise to gain a background in geology, chemistry, physics, engineering, or other technical subjects before taking courses specifically focused on the oil industry. People are also advised to gain practical experience while they are amassing university credentials. An ideal approach combines basic knowledge with technical training and on-the-job experience.

Table 3.1: Employment Trends in Selected Energy-Related Industries
(thousands of employees)

Year	Nonagricultural Establishments	Coal Mining	Oil and Gas Extraction	Petroleum Refining
1960	54189	186.1	309.2	177.2
1965	60765	141.4	287.1	148.1
1970	70880	145.1	270.1	153.7
1975	76945	212.7	328.8	152.8
1980	90406	246.3	559.7	154.8
1985	97519	187.3	582.9	141.4
1986	99525	175.9	450.5	131.1
1987	102200	161.8	401.8	125.2
1988	105336	150.8	400.7	120.8
1989	108413	145.9	385.7	118.1
1990	110321	154.6	408.4	119.9

Source: United States Department of Labor, Bureau of Labor Statistics.

Table 3.2: Historical Energy Consumption Patterns

Year	Annual Total Quads	Fuelwood Quads	%	Coal Quads	%	Petroleum and Natural Gas Quads	%	Hydropower Quads	%	Nuclear Quads	%
1860	3.1	2.6	83.5	0.5	16.4	—	—	—	—	—	—
1880	5.0	2.9	57.0	2.0	41.1	0.1	1.9	—	—	—	—
1900	9.6	2.0	21.1	6.8	71.3	0.5	5.0	0.3	2.6	—	—
1920	21.3	1.6	7.5	15.5	72.8	3.4	16.1	0.8	3.6	—	—
1940	25.0	1.4	5.3	12.5	50.1	10.2	40.9	0.9	3.7	—	—
1950	34.0	1.1	3.3	12.5	38.0	19.0	55.9	1.4	4.2	—	—
1960	44.6	—	—	10.1	22.8	32.8	73.5	1.7	3.7	—	—
1970	67.1	—	—	12.7	18.9	51.5	76.8	2.7	4.0	0.2	0.3
1975	70.6	—	—	12.8	18.2	52.6	74.6	3.2	4.6	1.8	2.6
1980	76.0	—	—	15.5	20.4	54.6	71.8	3.1	4.0	2.7	4.0
1989[1]	81.4	—	—	19.0	23.3	53.6	65.8	2.9	3.5	5.7	7.0

[1] 1989 data from *International Energy Annual*, U.S. Department of Energy.

Adapted from tables published by Bureau of the Mines; Bureau of the Census; Department of Labor.

OPPORTUNITIES FOR WOMEN

Much is being said about the problems women face when they seek jobs in what has been traditionally a "man's world," a vague term that can be associated easily with the sometimes dangerous, dirty, and physically demanding environment in which many energy-related jobs exist. State and federal laws are eliminating some of these problems and making progress toward alleviating others. Equal access to opportunities to advance to top level jobs in management seems to be one of the problems that continue to confront women.

There are few clear-cut suggestions on how women can improve their access to managerial positions. The media infer that the incentive to train for top management jobs is a factor and that there have been too few women who have held top jobs to create a pool of role models to inspire young women. Because they have been brought up in a society where girls have often received less encouragement to develop their business abilities, women have had difficulty building confidence in themselves.

Comments by two of Exxon's recruiters describe what they are looking for. They say that people being interviewed for jobs with Exxon are ranked on seven factors:

1. communication skills
2. interpersonal skills
3. sense of responsibility and commitment
4. aptitudes and abilities
5. job fit
6. other interests
7. overall evaluation

Each interview can be affected by the beliefs and feelings of both the interviewer and the applicant. One interviewer, commenting on the practice of hiring college students for summer jobs, says, "That's when we learn if a candidate is opinionated or just plain hard-headed. Sometimes that's a plus, sometimes a minus." Another interviewer says, "You've got to have a certain amount of gutsiness in this business. We look for grades; they usually tell you if the person is a hard worker. But I also ask myself, 'Is this person willing to go out on a limb in favor of a proposal?' After all, this whole business is built on risk.''

These comments indicate the qualities that are looked for in applicants. The job requirements for any applicant should be the same, regardless of sex, age, or racial or ethnic factors. Women, or members of minority groups, hold more key jobs today than they did ten, or even five, years ago. The only way they will get more of these jobs is to go ahead and get into the competition and win!

TRENDS IN EMPLOYMENT

Things that happened during the late 1970s and early 1980s teach us that there can be sharp changes in the number and kinds of job opportunities in the petroleum industry. In the first place, the price of crude oil is determined by a very complex set of factors. Similarly, the quantity of oil released to the market appears to be subject to decisions that are difficult to anticipate. Furthermore, how much gasoline and other petroleum products people buy can depend as much on the going price as on established usage patterns.

To illustrate this state of affairs, consider what happened to local service stations around 1980. Together with sharp changes in the price of gasoline, there were sharp increases in the cost of operating service stations, especially the cost of station attendants. In time, the self-serve idea became widely adopted and as a result, employment opportunities for station attendants decreased sharply.

Under similar circumstances, higher prices for petroleum products prompted people to consume smaller quantities. Prior to about 1981, the price of crude oil was high enough to prompt widespread drilling of new wells. Job opportunities in this part of the industry increased rapidly. Falling prices for oil, however, idled hundreds of drilling rigs with the obvious effect on employment opportunities. Take another look at Table 3.1.

Probably the best way to avoid being put to a disadvantage by unpredictable changes in employment opportunities lies in staying alert to conditions as they develop around any job. If things seem to be running against a particular kind of job, begin looking for related jobs for which the required training can be obtained. Gasoline service station operators or pump attendants who realize that high automobile prices would cause people to drive older cars protected their jobs by moving toward expanding their automobile repair facilities.

Rising crude oil prices are sure to provide incentive not only to search for new fields but also to extract a greater fraction of the oil that might have been left behind in existing fields.

The branch of the industry that is devoted to improving recovery from known fields is very likely to expand. In newly drilled wells the oil is often under so much pressure that it rises to the surface. In time, as the pressure decreases, the oil must be pumped to the surface. It is also necessary for the oil that is diffused throughout the total rock strata to find its way to the well. How efficiently the oil reaches the well depends upon the permeability of the rocks, the thickness or viscosity of the oil, and how much pressure is exerted by the overlying rock.

An expanding technology is being developed to increase the percentage of oil in a rock stratum that can be brought to the surface. The current average for recovering oil is about 32 percent. Obviously the remaining 68 percent poses an attractive challenge to the oil industry.

The simplest technique for enhancing oil recovery relies on drilling other wells in the vicinity of one or more producing wells and then pumping water down these injection wells into the rock strata where the oil comes from. Pumping water into these strata forces the oil to flow toward the producing wells. In practice, however, this method is not as good as it might appear to be. Not always are the underground rock formations as uniform as core samples might suggest, and this lack of uniformity can cause the oil to flow in unforeseen directions.

Sometimes, since oil and water don't mix, the water bypasses the oil. The water develops channels through which it flows, and the oil is left where it was. Much research is being aimed at finding substances that can be injected into oil-bearing rock strata that will either mix with the oil or otherwise solve the bypass problem. Of course these substances must be cheap and plentiful enough to allow the price of the recovered oil to compete in the marketplace.

In some oil-bearing formations the oil can be forced toward the producing wells by injecting carbon dioxide into the rock strata. Ordinarily, carbon dioxide under pressure and at low temperatures exists as a solid—that is, dry ice. Fortunately, the pressure and temperatures that exist in many oil-bearing rock formations allow carbon dioxide to exist as a liquid. Liquid carbon dioxide mixes with oil and can thus sweep the oil toward a producing well. It has the disadvantage of diffusing through rock strata more rapidly than does the oil, however, and this tends to leave too much of the oil behind.

Carbon dioxide is more expensive than water, but it is a by-product of several industries and is available in large quantities. This gas works well in oil fields that have been flooded with water. Many such wells are located in California, Texas, and in the states surrounding Pennsylvania. Based on exploratory efforts sponsored by the U.S. Department of Energy, it is estimated that, of the 100 billion barrels of oil remaining in these fields, carbon dioxide flooding may assist in producing 9 billion barrels of oil.

The Department of Energy is also sponsoring research in which steam is used to enhance the recovery of oil. The oil left behind in porous rocks flows more readily when heated, and steam can provide the heat. The problem is to keep the steam from losing its heat while it is being pumped down the well on its way to the oil-bearing strata. One solution being tested is to install steam generators at the bottom of the well.

It is noteworthy that the U.S. Department of Energy employs more than 3,000 people in energy-related careers and jobs. The administration of grants to support enhanced recovery of oil is only one of many programs in the department. People who are trained to hold jobs in the energy industry can also qualify for jobs in the DOE, and the number of these jobs is very likely to increase if the amount of energy available to our society becomes critical.

If we disregard the effects of sudden, short-term events such as the oil embargo of 1973, changes in the number of employment opportunities in the energy-related industries are likely to follow the overall industrial scene. See Table 3.1.

WHAT ABOUT OIL SHALE AND TAR SANDS?

A large number of highly challenging job and career opportunities exist in a 17,000 square mile area in Wyoming, Utah, Colorado, and central Canada. Some 30 to 60 million years ago these areas were covered with shallow lakes and swamps. By processes that are not yet understood, the organic material that collected in these areas as sediments has been changed to kerogen and other petroleum-like substances. Rock-forming processes together with disturbances in the earth's crustal formations have left the kerogen diffused in oil shales and tar sands.

Kerogen can be extracted from oil shale and converted to petroleum. The enormous quantities of this oil shale create an attractive challenge to all those looking for solutions to the problems that grow out of inadequate supplies of petroleum. The amount of oil that might be extracted from these huge shale oil formations is estimated to be three times greater than all of the oil in Saudi Arabia, or nearly twenty times greater than the amount of oil known to be in the rock formations under the United States.

In theory it is easy to extract oil from oil shale. When the shale is crushed and heated to about 900°F, the kerogen separates from the rock—first as a vapor mist, then as it cools, as a liquid much like crude oil.

We know that the dollar cost of recovering oil from shale creates economic problems. The cost in terms of the amount of energy that must be spent to get oil creates an even stickier problem. The production costs in energy terms must be less than the energy that can be made available when the shale oil–produced petroleum is burned.

Similar challenges exist in the even greater quantities of tar sands that are to be found in several western states and in Canada. These deposits contain an estimated 600 billion barrels of oil. Several oil companies have already built pilot plants to explore ways to extract oil from tar sands. In one project, wells were drilled into the tar sand strata. Steam was then forced into the sand strata to melt the tar so it could be pumped to the surface. Underground burning has also been tried as a way to melt the tar and allow it to be pumped to the surface. In either case, it must be pumped to the surface through heated pipes.

Skeletons of abandoned experimental plants to extract oil from tar sands bear witness to the difficulty of the problems that must be solved before this energy resource can be used to fill our gasoline tanks and fire our oil-burning furnaces.

For sure, many jobs and career opportunities await people in the location of new oil fields, in the production and distribution of oil, in improving methods to extract a greater fraction of the oil from underground rock formations, and in increasing the efficiency of refinery operations. We can describe less confidently the job and career opportunities that are developing in those phases of our society where our nation's oil supply becomes involved in economics and political activities. The seriousness of this involvement, however, suggests that there will be a great need for people who understand the nature of petroleum as an energy resource, who can evaluate estimates of our supply and demand, and who can foresee how governmental regulations and international relations affect and are affected by the problems that arise when demands exceed supplies.

OIL INDUSTRY JOBS AT A GLANCE

Engineers and Scientists
(Generally requires a four-year bachelor's degree, but may require postgraduate study)

petroleum engineers	chemists
environmental engineers	geologists
chemical engineers	geophysicists

Support Workers
(Generally requires at least two-year associate degree for technicians; technologists require a four-year bachelor's degree)

engineering technicians laboratory technicians
engineering technologists computer operators
computer programmers

Blue Collar Workers
(Generally requires an apprenticeship)

welders oil well drillers
ironworkers

Semiskilled Workers
(May require a high school education)

truck drivers laborers (roughnecks)
equipment operators

CHAPTER 4

CAREERS IN THE COAL INDUSTRY

The role that coal has played in providing us with energy during the past 100 years brings home a valuable lesson for people who are making career plans: As the demand for an energy resource rises or falls, so will the number of job and career opportunities that are associated with that resource.

Although there is a relatively abundant supply of coal in the United States, the demand for this energy resource can rise and fall in response to changes in the business and industrial world. About 100 years ago, coal displaced wood as the leading energy resource. It held this position until about 1940 when oil became the nation's leading source of energy with natural gas running a close second. Since 1940 coal has provided between one-fourth and one-fifth of the nation's energy.

Keeping in mind the total supplies of coal, gas, and oil and taking a long-range view, there is the possibility that the diminishing supply of oil and natural gas, especially within the United States, will force another sharp shift in the kind of energy resource that will produce the greatest number of jobs within the next ten years or so. The demand for coal will always depend on whether other energy sources—such as natural gas and nuclear energy—are available at reasonable cost.

Employment opportunities will fluctuate with the demand for coal, and are consequently difficult to predict. However, the Department of Labor is not overly optimistic. For example, job prospects for mining engineers are expected to be below average through the year 2000. Although, as technology becomes more sophisticated and environmental, health, and safety concerns present new challenges, there will always be a place for talented newcomers in the industry. Ironically, these same technological advances may mean fewer jobs for those who mine coal as techniques become more efficient and less labor-intensive.

THE NATURE OF THIS ENERGY RESOURCE

Coal beds exist today where dense prehistoric swampy forests once thrived. Changes in the earth's climate extending over millions of years have been accompanied by alternate sinkings and risings of large sections of whole continents. When a sinking crustal segment became sufficiently submerged, sediment would collect from the erosion of emerged areas. This submerging of swampy areas led to the gradual burying with sediments of the plant material that had collected in the swamp.

In the Everglades of South Florida there is a blanket of water-saturated peat averaging more than six feet in depth. The ocean lies to the southwest and land to the northeast. So long as the region continues to subside gently, the thickness of the peat deposit will increase. If the rate of subsidence increases, however, then the peat deposit will be covered over by limestone-forming marine sediments or the sands and shales being eroded from the land areas to the northeast. The pressure of the accumulating sediment aids in converting the peat to coal.

Many of the coal fields of North America occur in rock formations containing alternating sequences of marine and landmass sediments. Sometimes as many as twelve layers of coal are sandwiched between beds of limestone, shale, and sandstone. Apparently, when the plant debris from a submerging swamp was buried at greater and greater depths, the hydrogen and oxygen in the carbohydrates of the plant tissues was distilled off leaving the carbon residues behind.

There are coal beds in areas of the world where today's climate supports practically no plant growth. This fact bears witness to the amazing changes that have occurred in large areas of the earth's surface. For example, thick coal beds exist in arctic regions where the surface is now covered with layers of ice and snow. Supposedly, during sunnier days, swamps flourished in what are now arctic areas. Equally dramatic changes in climate have occurred in many parts of the world and layers of "buried sunshine" lie in the rock strata beneath the surface.

Chemical Makeup of Coal

Coal is a black, solid mass consisting of carbon, hydrogen, and oxygen. The carbon content determines whether a particular deposit is peat, lignite, bituminous, or anthracite. As the carbon content increases, so does the hardness. Some forms of anthracite are almost pure carbon and are so hard that they will scarcely burn.

Coal contains varying amounts of sulfur, which burns to form sulfur dioxide, the chief pollutant released by the burning coal. Generally, the coals of eastern North America are high-sulfur coals (over 3 percent), whereas coal from western North America contains little sulfur.

HISTORY OF THE COAL INDUSTRY

Sometimes deeply buried coal beds are exposed along the banks of streams or brought to the surface during mountain-building or other geological events. Primitive people may have used the black rocks that fell from exposed coal beds to build their cooking fireplaces, thereby "discovering" coal. Archaeological evidence suggests that coal was used in funeral pyres as many as three thousand years ago. Metal workers a thousand years before the Christian era may have used coal to fuel their Bronze Age forges. Early Greek history and the Bible provide additional evidence that coal does, indeed, have a long history.

The Middle Ages and Renaissance

By the 1300s, so many Londoners were burning the coal that could be picked up along England's coasts that Edward I outlawed its use. He wanted no part of a fuel that filled the air with unsightly smoke and bad smells. It took several hundred years for people to invent ways to burn coal cleanly enough to enable it to compete with wood and charcoal; but the British were burning two million tons of coal per year by 1660 and nearly eight million tons annually by 1760.

The Industrial Revolution

Although coal was not mined in America until 1760, the Indians knew where to find coal and how to use it to make pottery. By 1860, the per capita use of coal was not much more than thirty pounds per year.

The impact of the Industrial Revolution of the late 1800s caused the use of coal to increase rapidly, and by 1900 some 270 million tons of coal were being burned each year.

The development of the coal industry created profound changes in the way of life in the United States. Before industry expanded during the Industrial Revolution, people spent most of their time growing food, weaving cloth, making bricks, processing lumber, winning metals from ores, fashioning tools, and doing all of the kinds of hand labor needed to keep their communities going.

Learning to mine coal and to use its energy to power the world's engines is tied closely to the origins of the Industrial Revolution. Changes in the working world were sure to occur when one person could dig enough coal to replace the muscle power of hundreds of workers or dozens of horses. In contrast to wind and waterpower, which were dependent upon weather conditions, coal was stockpiled easily; industry could plan ahead for increasingly greater production.

As industry expanded, farmers left their farms and craftworkers left their family enterprises and took jobs in the increasingly mechanized mines and factories. People became less inclined to grow their own food or make their own clothes, furniture, and tools; they relied on wages to buy the things they needed. There is a world of difference between life-styles built around tending farms or working in cottage industries in country settings and being a small part of a large factory or multinational enterprise in an urban environment.

Use of coal continued to increase rapidly throughout the 1900s. By 1990, close to 900 million tons were being burned each year.

There is an estimated three-trillion-ton supply of coal in the United States and a whole earth supply of well above five trillion tons. These estimates, together with the known rate of coal consumption, suggest that there is enough coal to supply human needs for hundreds of years.

TYPES OF COAL MINES

Each coal mine begins with the location of an underground bed of coal. If the coal bed is thick enough, and if the bed extends over an area wide enough to indicate that it would yield commercial quantities of coal, samples are taken and tested for such properties as carbon, ash, and sulfur content and for general fuel value. When data suggest that several million tons of marketable coal await being harvested, the area of the potential mine is carefully mapped. Somewhere along the way, the coal company either purchases the land or arranges with the owner for the right to mine and market the coal.

To be profitable under existing economic and technological conditions, the coat bed should be at least three feet thick. Coal beds are rarely more than eight feet thick, but occasionally beds more than forty feet thick are discovered. Coal beds are always sandwiched between layers of shale, sandstone, or other kinds of rock or clay. The beds may be as level as they were when they were first laid down at the bottom of a swamp, or movements of the earth's crust may have left them sloping, tipped, or warped into irregular shapes. They may be exposed on hillsides or along stream banks or they may be completely hidden under layers of rock hundreds of feet thick.

The character of a coal mine is determined by the location and condition of the coal bed to be mined. If the bed is exposed on a hillside, a drift type of mine simply follows the layer of coal deeper into the hill. In areas where the coal beds are tilted, the mines take the form of a sloping tunnel.

Underground mines must be provided with ventilating systems to bring fresh air to the miners and remove harmful and sometimes explosive gases that may collect in the mine tunnels. Water tends to seep into most mines, and this water must be collected and pumped to the surface.

MINING PROCESSES

Underground Mining

In modern underground mines, specially designed machines called continuous miners break the coal loose and transfer it to the surface. In less mechanized mines, the face of the coal bed is attacked with huge chain saws that cut slots into the coal or with drilling machines that bore deep holes. Explosives or cylinders of compressed air are used to blast the coal loose. Loading machines then scoop and sweep the loosened coal into low-slung, electrically driven trucks or onto conveyor belts.

Designing machines that can be used in the cramped conditions of a mine creates a special challenge to engineers and designers. In contrast to the usual above-ground jobs where we see bulldozers, dump trucks, and other heavy equipment handling rocks, all mine machinery must be designed to do the same kinds of work but be no more than six or eight feet tall.

Heading off accidents is another challenge to the people who design and operate mining machinery and supervise mining operations. In all types of mines, precautions must be taken to keep the roof from falling in as the coal is removed. Carefully considered decisions determine how large to make the pillars that are left to support the roof. For additional safety, as rapidly as the coal is taken from the working face of the mine, holes are driven through the overhead layers of rocks and steel rods inserted to bind them together for added support.

Cave-ins, rockfalls, and explosions are constant threats, and adequate safety precautions are recognized as an essential part of mining. During some years, when compared with all other industries, the mining industry shows the lowest number of accidents per million worker-hours.

In today's most efficient underground mines, amazingly adaptable machines move back and forth along the face of a coal deposit, chew off chunks of the coal, and drop it onto conveyors that take the coal to the surface. These longwall mining machines are fitted with a movable steel canopy to support the roof over the mined-out coal. As the mining machine moves into the wall of coal, however, the roof is permitted to fall behind it in a controlled and safe manner.

Strip Mining

Where the coal beds are relatively near the surface, the coal is removed by strip mining. In this type of mine, giant power shovels, drag lines, or other earth-moving equipment move aside the overburden of rocks and soil. Once the coal is laid bare, smaller power shovels break the coal loose and load it onto trucks or conveyors.

Surface mining, to be profitable, usually requires a minimum seam thickness of five feet, assuming that the amount of overburden that has to be removed is not excessive. Where several layers of coal can be recovered at the same strip mine, profitable operations permit as much as several hundred feet of overburden to be removed. A total seam thickness of forty feet, for example, would justify removal of up to five hundred feet of overburden. For underground mines, the usual minimum seam thickness is five feet and at a depth not to exceed three thousand feet.

Designed to be as efficient as possible, strip mining uses some of the largest machines in the world. Shovels can be more than two hundred feet tall and weigh millions of pounds. The dippers can take as much as 270 tons of rock at one bite and drop it more than a city block away. Modern machines can harvest coal by the surface mining method that a few years ago would have had to be mined by underground operations.

In 1940, less than 10 percent of the coal burned in the United States was strip-mined; today more than 40 percent is obtained by this method. This trend points toward changes in the kinds of job opportunities that exist in the coal industry. In addition to the jobs that are involved in getting the coal from the ground and on its way to market, many jobs involve restoring the mined area to a condition that will support agriculture, recreation, or other worthy land use.

Damage to the countryside has been portrayed vividly by environmentalists whose jobs require them to seek legislation that will insure the reclamation of strip-mined land. Representatives of the coal companies must be able to estimate the cost of restoring the land to make new land use possible. Obviously, unless these costs are included in the total cost of producing and marketing the coal, the expense of reclaiming the land must be borne by other agencies.

Reclaiming land that was strip-mined by coal companies that are no longer in business is creating challenging job opportunities. These opportunities range all the way from public relations jobs that arouse public interest or generate appropriate legislation to seeking new species of plants that can survive in the sometimes hostile environment of newly reclaimed land. There are special challenges for landscape architects and land use planning experts who try to work out the best locations for parks, lakes, campsites, housing developments, or other enterprises that would be beneficial to an entire community.

KINDS AND NUMBERS OF JOB OPPORTUNITIES

Some 165,000 people find jobs in the coal industry. The titles of many of their jobs reflect what goes on where coal is mined, processed, transported, and marketed.

There are drilling machine operators, cutting machine operators, loading machine operators, and continuous-mining machine operators. There are shot-firers, roof bolters, and stopping builders who build partitions in the mine passageways so the movement of air can be controlled.

Fire bosses and safety engineers inspect the working area and make sure there is minimum danger from failing rocks, dangerous gases, or inadequate ventilation. Rock-dust machine operators spray ground limestone on the mine walls to reduce the amount of coal dust in the air.

Similar job titles describe the work done in strip mines, although the machinery differs. For example, rotary auger operators use giant augers that bore into the face of a layer of coal if it is beneath such a thick layer of overburden that it would not be economical to strip it away.

There are jobs for people who install and repair machinery, check electrical systems, and build whatever is needed to operate the mines. The skills are pretty much the same as those for people who do similar work above ground, but the working conditions are unique. Repair people sometimes find themselves working on their knees with only their headlamps to provide light until the electrical system has been restored.

EDUCATION AND TRAINING

"Coal: The 500 Year Career," a publication of the West Virginia Coal Association, suggests that people who are interested in qualifying for a job or career in the coal industry ask themselves three questions:

1. Where are you?
2. What can you do?
3. Where can you be?

Here are the kinds of answers the publication anticipates: I'm in school . . . but not yet a high school senior. I'm a senior in high school interested in training or advanced education. I'm in high school now and have a serious interest in mine engineering. I'm a high school graduate who would like to enter the coal mining industry as soon as possible. I'm out of school, either a graduate or nongraduate, and would like to know if my present skills would be of greater reward to me in the coal industry.

Having answered the three questions, young people are advised to talk their answers through with school counselors and to seek information on special training programs offered by government agencies in cooperation with the coal industry. Where appropriate, would-be employees are advised to visit the personnel directors of nearby mines. Some coal companies sponsor work-study programs that enable young people to earn wages while they are continuing their engineering or other technical training.

Summer jobs are also cited as providing excellent opportunities to be introduced to the entire mining industry. For people who are employed in other fields, schools and colleges in mine communities sometimes offer adult education evening programs that build the specific technical skills needed in the coal industry.

Most miners start out as helpers to experienced workers and move to higher paying jobs as they gain experience. It is general practice to post notices of job openings, and all qualified workers may bid for jobs that would represent career advancement. Jobs are filled on the basis of seniority and ability. People can begin as helpers and by gaining experience in the mine and completing training programs sponsored by the industry. They can then advance to supervisory and administrative positions in the company.

TRENDS IN EMPLOYMENT

There are definite indications that the demand for coal will increase as the price and availability of oil and gas put cost pressure on these two energy resources. Greater use of coal is also predicted as the industry solves the environmental problems associated with strip mining and with air pollution from burning high sulfur content coal.

Highly promising career opportunities exist in the research and development efforts to make artificial gas and liquid fuels from coal.

Coal, oil, and gas are products of the same general chemical processes. All of these fuels can be traced to partially oxidized and chemically altered plant remains. They are all made of carbon and a few other elements, mostly hydrogen and oxygen.

The main differences between gas, oil, and coal are differences in the ratio of carbon to the other elements. In coal there are about equal numbers of carbon and hydrogen atoms; in oil, about two hydrogen atoms for each carbon atom; and in gas, four hydrogen atoms for each carbon atom.

One process for converting coal to gaseous fuel dates back to the early 1900s before natural gas became available. Pulverized coal, oxygen, and steam are brought together in a gasifier. By introducing the pulverized coal and the other gases at opposite ends of the gasifier and by keeping the gaseous mixture heated to about 3,500°F, the coal is gasified almost instantly and completely.

This gasified coal is a mixture of carbon monoxide and hydrogen and yields about 75 percent of the energy originally present in the coal. The gas is washed as it leaves the gasifier. This removes bits of coal ash. If the coal contains sulfur, bad smelling hydrogen sulfide may be produced and must be removed by absorbent systems.

For more than a hundred years people have been trying to obtain gaseous fuel from coal while it is still underground. A patent on underground coal gasification was granted in 1909, but the first large-scale development of this idea took place in the Soviet Union in 1933. Large-scale exploratory work has also been done in the United States, but serious difficulties have kept the idea from catching on.

All of the methods tried so far require the underground coal bed to be at least partially fractured or opened. This is done by explosives or by boring holes or cutting shafts into the coal bed. Air or pure oxygen is then forced into the fractured coal to convert the coal to a gaseous fuel. All methods must be controlled carefully to insure that the coal is not burned all the way to carbon dioxide. Since the volume of the reacting "container" is always changing, so too is the ratio of the reacting materials. To change the "recipe" is likely to change the finished product.

Ways to convert the energy in coal directly to electrical energy are the goals of additional research. The idea is to bypass the usual steps whereby coal is burned to heat water which forms steam which drives turbines which spin generators to create electricity. Pilot plants that use coal-fueled fuel cells are already operating.

Magnetohydrodynamics technology, as it is called, is also being tried as a way to convert coal energy directly to electrical energy. In this method the coal is used to produce high-pressure, high-temperature gases which are passed through a strong magnetic field. Since the stream of gases consists of electrically charged particles, electrical energy is generated when the gas stream cuts the magnetic field.

COAL AND ENVIRONMENTAL POLLUTION

Writing in *Fortune* magazine, Tom Alexander discusses several problems that are linked with coal as an energy resource. The title of his article, "New Fears Surround the Shift to Coal," tips us off to the main emphasis of the piece. He spells out how the increased use of coal is sure to call for greater efforts to prevent environmental pollution.

Careers in the energy industry will become increasingly linked with careers that involve preventing environmental pollution.

Many people are already devoting their lives to efforts to improve the quality of our environment. Some of these jobs take the form of creating programs to arouse public concern for the ill effects of messing up the environment. Environmentally-related illnesses such as the "black lung" disease of coal miners have been given widespread publicity. A national program called Earth Day has triggered many highly dramatic efforts to make people aware of the various causes of environmental pollution.

An increasing number of significant job and career opportunities await people who are trained to head off environmental pollution at its sources.

As Alexander brings out in the *Fortune* article, coal mining has always involved severe threats to the quality of the environment.

Spoil piles linger long after mines have been abandoned. Acid drainage from abandoned mines continues to befoul streams even after vegetation has covered up all evidence of abandoned mining operations. For years urban death rates were higher than those in the country, and the principal reason has been assumed to be the noxious emissions from industrial and domestic coal-burning furnaces.

It is the phrase *assumed to be* that points most directly to the career opportunities that will receive added emphasis in the future. Much research needs to be done to spell out exactly how one or another alleged source of pollution messes up the environment. For example, Alexander cites the case of blaming sulfur dioxide for many of the harmful effects from coal smoke. In one experiment, laboratory animals were made to breathe sulfur dioxide at concentrations much higher than those in the air in badly polluted neighborhoods. Contrary to expectations, the animals showed no apparent harmful effects. Obviously, additional research is needed in this area.

Millions of dollars are being spent to equip coal-burning installations with expensive filters known as "baghouses" to remove particles from the coal smoke. Equally expensive "scrubbers" are installed to wash out the sulfur dioxide. Obviously, these installations help to alleviate the environmental pollution problem. Future research, however, may point to even more effective measures if the nation is called upon to turn to coal for a larger proportion of its energy supply.

People who are trained both in energy production and utilization and who, at the same time, are aware of and knowledgeable in the scientific principles involved in environmental pollution may find worthwhile careers awaiting them.

COAL INDUSTRY JOBS AT A GLANCE

Engineers and Scientists
(Generally requires a four-year bachelor's degree and may require additional postgraduate study)

mining engineers analytical chemists

civil engineers   environmental engineers
geologists   mine safety inspectors
geophysicists   coal gasification engineers

Support Workers
(Generally requires at least a two-year associate
degree for technicians; technologists require a
four-year bachelor's degree)
engineering technicians   laboratory technicians
engineering technologists

Blue Collar Workers
(Generally requires on-the-job training or an
apprenticeship following a high school education)
mine workers (drilling machine operators; roof bolters;
etc.)

Semiskilled Workers
(May require high school education)
miner's assistants   truck drivers
laborers

CHAPTER 5

CAREERS IN THE NATURAL GAS INDUSTRY

THE HISTORY OF THE GAS INDUSTRY

The story of how the natural gas industry has developed tells us much about how the world's people learn to take advantage of the earth's energy resources. During America's infancy, natural gas was not much more than a puzzling curiosity. In several parts of the world, "burning springs" added a ghostly glow to the nighttime landscape. Night after night, blue and yellow flames would flicker among the rocks and ripples of these mysterious springs. We have been told that George Washington was so fascinated by a burning spring he found in the Kanawha River Valley near present-day Charleston, West Virginia, that he purchased the surrounding land.

Many years passed, however, before people learned how to take advantage of this natural energy resource. During the 1820s, an inventive young man, William A. Hart, managed to use the gas from a burning spring in the northwest corner of New York State to replace candles and oil lamps. He used hollowed-out logs to capture the gas before it ignited or escaped into the air and to carry the gas to where he wanted to use it. His success is reported in newspaper stories that say that several buildings in Fredonia, New York, were lit with gas as early as 1825. Hart also convinced the United States government to replace the oil-fueled lamp in a nearby lighthouse on Lake Erie with a gas-fueled lamp. Gas that was transported through wood pipes kept the lamp burning for more than twenty-five years.

The gas-burning lighthouse is now a tourist attraction and a memorial to the pioneering spirit of William A. Hart, the first American to enjoy a career in the gas industry. Today, some fifty million homes in the United States are cooking, heating, or lighting the gas. So are some three million commercial and industrial buildings. Throughout the total global community, more than sixty trillion cubic feet of gas are used each year as a fuel or as a raw material for the chemical industry.

The development of the gas industry brings out many of the ups and downs, ins and outs of the total story of people learning to utilize, manage, and conserve the earth's energy resources. Particularly noteworthy is how the development of an energy resource is tied in with other technological, economic and sociopolitical developments. One quite revealing example was the dependence of the gas industry on the iron and steel industry's learning to make steel pipes that wouldn't leak and were large enough, strong enough, and cheap enough to solve the transport problem. It took nearly one hundred years before steel pipe could replace Hart's wood pipes.

As soon as the transport problem was solved, the gas industry boomed. And it continued to do so well into the 1900s with an obvious effect on career opportunities. In fact, it is easy to see how solving the transport problem created increasing career opportunities in the gas industry, at least in retrospect. In sharp contrast, things can happen that make it exceedingly difficult to anticipate changes in the availability of job opportunities in all of the fossil fuel industries.

Before the 1970s, few people were concerned lest the earth's supply of the fossil fuels might be less than infinite. Then came the massive impact of the fuel shortage which plagued the United States during the early 1970s. Inadequate supplies created long lines at gasoline service stations. Builders and new home owners were told that local gas companies could not take on new customers, nor could industries and power-generating systems obtain adequate supplies. Degrees of daily inconvenience were matched only by fear lest the nation's future and its citizens' life-styles be forever circumscribed by inadequate supplies of oil and gas.

ESTIMATING THE WORLD SUPPLY OF ENERGY RESOURCES

All of which is to say that people who are looking forward to energy-related careers have a large stake in efforts to estimate how much gas, oil, and coal there is in the world and how rapidly the supply is being diminished. To protect this stake will not be easy. On the positive side, several agencies are striving to come up with truly valid estimates. At the same time, it is to the advantage of several vested interests to proclaim the possibility of immediate exhaustion of the earth's supply of oil and gas or, in complete contradiction, to proclaim equally strongly that humanity's demands for these resources will never exceed the supply.

Trying to describe this controversial state of affairs by analogy risks being more misleading than enlightening, more disconcerting than satisfying. But keeping in mind the origin of the earth's store of fossil fuels, it may be appropriate to think of these resources as being analogous to goodies that have been hidden in an earth-sized layered birthday cake. But this cake did not arrive in good condition. On the contrary, it is as though a delivery system dropped it from a great height and it now exists as a badly jumbled mass in which the original layers are scarcely evident. There is no easy way to determine how much of each prize was initially present or has been retained in any one of the shattered chunks.

Furthermore, one of the hidden prizes is volatile (gas), another is more or less liquid (oil), and a third is solid (coal). How much of each prize remains in any one chunk depends upon circumstances that are pretty much up to chance.

Continuing the analogy, the guests at this birthday party introduce additional difficulties. Appetites may differ sharply and so may attitudes regarding conservation. Problems arise because all chunks of this birthday cake do not contain equal portions of the hidden goodies, and there are sure to be conflicts over who actually owns the best chunks. In fact, behavior at this children's party may have its counterpart among nations in the global community if people sense that the earth's supply of fossil fuels may be running out.

In more appropriate language, "Welcome to uncertainty" is the phrase used professionally "to introduce both laymen and professionals to the maze of petroleum energy data that must be comprehended to achieve understanding of this critical commodity." Furthermore, Masters and his colleagues at the U.S. Geological Survey point out, "we have made progress in removing uncertainty from energy data, but, in general, we must accept that there are many points of view and many ways for the 'blindman to describe the elephant.' "

At the same time, Masters sees "broad agreement on the dimension of petroleum resources" and finds that "assessments of natural gas, as for oil, have shown a remarkable consistency."

Based on U.S. Geological Survey estimates, both the United States and Canada will be actively seeking new sources of energy early in the 2000s, but Mexico will not be under the same urgency for nearly thirty years.

Adding optimism to career opportunities in the gas industry is the fact that gas can be made from coal. If water is sprayed onto a bed of hot, burning coal, hydrogen and carbon monoxide are produced. If the mixture of these gases is passed over a nickel catalyst while they are still hot, a new reaction yields methane and water. When these hot gases cool, the water condenses and can be drained away, leaving the methane to be sold as artificial rather than natural gas.

Processes for converting goal to gas are being improved. In one improved method, the coal is ground to a fine powder. A blast of hot air mixes the coal with fuel oil to form a thin mud or slurry. The slurry, in turn, is sprayed into a chamber that contains a bed of burning coal. The gases that are produced contain methane.

CAREER OPPORTUNITIES IN THE GAS INDUSTRY

The gas industry offers a wide variety of career opportunities. Distribution companies provide widely available jobs. These companies purchase gas from transmission companies and pipe it to their customers. Transmission companies operate pumping stations that transport gas through a network of pipelines that link gas fields with customers. Production companies look for new sources of gas and drill wells. Somewhere along the way, impurities must be removed and, for most users, a substance with an identifiable odor is mixed with the odorless natural gas so leaks can be detected.

An American Gas Association career booklet emphasizes that the gas industry employs people with all kinds of backgrounds, skills, and training. For many jobs, a four-year-college degree is a minimum requirement. Other jobs may require a two-year college degree or a high school diploma. Vocational training and work experience often give job applicants an advantage. It is interesting to see that the skills needed for positions ranging from accountant to engineer include creative thinking and the ability to communicate and help others in a friendly manner, to work with a minimum of supervision, and to get along well with others.

CHAPTER 6

CAREERS IN THE SOLAR ENERGY INDUSTRY

Careers in the solar energy field are particularly attractive to people who are as concerned about the future as they are of the past or present. For sure, there are good reasons to take pride in making adequate supplies of energy available for our contemporaries, but there are also good reasons to keep in mind the equal dependence of future generations on nature's energy resources. If we are to play fair with our children and grandchildren, we are obliged to build into our utilization of energy the ways and means whereby adequate supplies will be available for those who come after us.

It is noteworthy that the careers of William Avery, Peter Glaser, and Sherwood Idso, the three people whose careers were mentioned in Chapter 1 to introduce careers in the energy field, were very much future-oriented. It is not by chance that solar energy plays the leading role in their careers. When all things are taken into account, solar and nuclear energy are sure to continue to be the fundamental sources of humanity's energy as far into the future as our imaginations can take us. Only geothermal energy resources supplement solar and nuclear.

UNDERSTANDING SOLAR ENERGY

The story of solar energy is very much a case of "first the good news and then the bad." The good news is the enormous amount of energy that the earth receives, and there are no good reasons to worry about the depletion of this resource for as far into the future as we can see. Many people have tried to translate its mind-boggling dimensions into familiar perspective. Information that was put together during the 1970s by Kenneth L. Laws and reported in *The Science Teacher* also touches on the "good news" side of solar energy. He calculated that the amount of energy falling on Carlisle, Pennsylvania—his home town—is almost one hundred times the town's average electrical power needs. He also calculated the amount of energy that is involved in three other natural energy resources that are related to solar energy.

Assuming that a twelve mile per hour wind were to blow constantly through Carlisle, the kinetic energy of the wind would be about equal to the town's electrical energy needs. If a ten degree drop occurs in the temperature of the air passing over the town, the energy given up would be about seven times the electrical power requirements of the town for one day. Finally, assuming that about thirty-seven inches of rain falls each year on Carlisle, the amount of energy released when this quantity of water condensed from the vapor state would furnish forty times the electricity used by the town.

Many people have made similar efforts to help us grasp the fabulously large amount of energy that comes to us each day from the sun. For example, the solar energy that falls each day on the continental United States is equivalent to almost four times the energy derived from burning a whole year's supply of oil. Another staggering fact is that in a period of less than two weeks, the earth receives from the sun an amount of energy equal to the earth's total known reserves of coal, gas, oil, and uranium.

Each year the sun sends toward the earth thirty thousand times as much energy as is used by the entire world's industries. Or five thousand times as much energy as is stored in all of the earth's volcanoes, hot springs, and similar geothermal phenomena. Or sixty thousand times as much energy as is spent keeping the oceans' tides ebbing and flowing.

When we realize the enormous amount of solar energy that flows to the earth from a source that is practically inexhaustible, we can see why people are specially attracted to the career opportunities that are tied in with solar energy. Before deciding to pursue these careers, however, it is wise to think about the problems that are associated with the commercial exploitation of solar energy. In other words, now for the bad news.

There are two big problems: First, solar energy comes to us in very diluted form. Second, not only is it turned on and off with the cycle of day and night, but the amount of solar energy we can harvest each day depends upon the season of the year and the weather at the moment.

The fledgling solar energy industry calls for people who can look at the business of developing an energy resource from a point of view quite unlike that of most established energy industries. Marketing solar energy is not a matter of finding and gaining control of a large amount of energy stored in a small "package"—as with oil and gas fields or coal and uranium mines—and then distributing the energy resource to customers. Solar energy comes to us already distributed. The problem is to gather it, concentrate and store it, and then make it available to customers when they need it in whatever quantities they are willing to pay for.

THE NATURE OF THIS ENERGY RESOURCE

At the risk of conflicting with the true nature of energy, think of the solar energy that falls on the earth as being like rain. We know that rain can be collected, stored, and then redistributed when it is needed and in whatever quantities people are willing to buy. To "bottle" sunshine, however, so that it can be delivered night and day, year in and year out, is a much more difficult operation.

Solar or any other form of energy can be "bottled up" only by allowing the energy to interact with some form of matter. For example, solar energy can interact with water and the energy remains in the water as heat. And as every photographer knows, solar energy can react with the sensitive emulsion on photographic film and be "stored" in the form of a latent image waiting to be developed in the darkroom. We know that solar energy falling on fields of corn, soybeans, potatoes, and dozens of other food crops can be "stored" in the form of energy-rich carbohydrates, fats, and proteins.

In keeping with the inescapable Second Law of Thermodynamics, each time a quantity of energy is changed from one to another form, a fraction of the energy is lost. It is this irretrievable loss (as discussed in Chapter 2) that creates much of the "bad news" side of the solar energy story.

The story of solar energy today and the challenge of the solar energy industry in the future depends upon finding ways to improve the efficiency of the energy transformations that stand between sunshine and whatever form of energy is needed to meet customers' demands. Here are to be found large numbers of potentially rewarding job and career opportunities.

Companies in Australia, the United States, England, and several other countries are marketing solar water heating systems for homes and commercial buildings. Each newly installed set of solar energy collecting panels bears witness to the number of job opportunities being created by this new industry.

Energy storage problems must be faced head-on in the design of solar heating systems for homes or commercial buildings. Architects are needed who can design buildings that will collect and store energy efficiently. Where buildings are located on their sites is also important. Giving each building its own "spot in the sun" without having to destroy trees or interfere with neighborhood buildings calls for careful planning. If taken into account early on in the design of a building, the required collecting plates can be installed without compromising the symmetry and general attractiveness of a building.

Architects who specialize in designing solar-heated buildings have one thing going for them: They know they have an energy resource that can be relied upon to deliver energy regularly no matter what kinds of political, economic, or embargo conditions develop.

People have always used solar energy to warm their homes and to raise their food. The oldest large-scale industrial use of solar energy, however, is to evaporate water from salt brines. The goal is either to produce dry salt or to recondense the evaporated water and thereby convert saltwater to freshwater.

It is easy to see why this is an early step along the way to exploiting solar energy. The technology bypasses the problem of storing energy. There is only one energy transformation involved—changing sunlight to heat energy. Nor is there a need to concentrate the solar energy so as to achieve higher temperatures than those that are produced by normal sunlight or to accommodate short-lived demands for large energy output.

Building solar energy cookers is another step along the way to a solar energy industry. Specially designed lenses or curved mirrors collect and concentrate the sun's rays so that cooking temperatures are achieved. The design of solar-powered water heaters began with simply arranging to have the water to be heated soak up the heat from the sun. More sophisticated versions use a special fluid to absorb the solar energy and then transfer the energy to the water to be heated.

Modern theories say that the sun's heat is produced by interactions between the nuclei of atoms within the sun. Supposedly, nuclei of the smallest kinds of atoms crash into each other and fuse to become nuclei of heavier atoms. During this process, a fraction of the mass of the lighter atoms is converted to energy. More is said about this kind of nuclear reaction later. For now, the theories about the origin of the sun's energy are well documented and support the belief that the sun will continue to send energy toward the earth for another thirty billion years. This is why we say that the sun is a finite but inexhaustible energy resource.

THE SOLAR ENERGY INDUSTRY

The solar energy industry appeals to a unique sort of person. People who look to this industry for career opportunities seem to be intrigued by the enormous potential—the fantastic quantities of energy that are available. They are equally intrigued by the market value of this energy and what it can do for the well-being of humanity. It is easy for these people to dream of the day that the world's energy supply problems will be solved by harnessing the sun's inexhaustible energy resources.

But these people also know that they must face the cold hard facts wrapped up in the laws that describe the way energy behaves. The laws of physics describe the unavoidable restrictions that are placed on any efforts to collect, store, transform, and distribute energy efficiently.

We are talking here about two kinds of efficiency. First, solar energy can only be developed at a cost that competes favorably with other sources of energy. Second, facilities for exploiting solar energy cannot consume more energy than these facilities will deliver to customers. To be successful in jobs in the emerging solar energy industry, people will need to be able to think in terms of both dollar accountability and total energy loss or gain.

Some branches of the solar energy industry are cost competitive today. Other branches could easily become cost competitive if the cost of coal, oil, and gas continues to increase. And then there are ideas under development that cannot become cost competitive until difficult technological problems are solved.

The so-called passive solar systems are totally cost competitive. In these systems, solar energy is transformed directly to heat energy. Buildings are being designed to take maximum advantage of the sun's energy to produce heat in winter and to provide cooling and ventilation in the summer. Passive solar energy design is typified by large areas of southernly exposed glass and energy-conserving insulation.

An early contribution to passive solar energy design was described in the *American Vocational Journal* for January 1978. Advanced vocational students who were enrolled in the building trades program at Zion-Benton Township High School in Zion, Illinois, pooled their homework and on-the-job training and built a solar energy house. Their designs used both passive and active solar heating.

The passive phase of this heating is apparent in the following quotation: "All windows will be fabricated from clear-plate, triple-insulating glass. The window position and size have been calculated to receive direct winter solar gains while shutting out the sun's rays in summer. The glass area is limited; yet it provides visual openness and natural illumination. Three-foot overhangs provide window shading during the summer, but because of the lower angle of the sun, will not shut out the sun's rays in the wintertime.

"The entrance door on the north side of the building and a sliding glass door on the south side are thermally insulated to further reduce heat loss. All entrance areas are recessed.

"The kitchen and laundry areas, which generate internal heat, act as living area buffers. Since they are not occupied continuously and can remain cool when not in use, they are located on the north wall. The living and family rooms are located on the south side. Here the sun will radiate warmth and provide natural illumination for these areas which are in constant use."

The students at Zion-Benton High School also installed an active solar heating system in the house. A 2,100-square-foot collector was set at a 55-degree angle on the roof. The roof had been designed to allow optimum collection of the sun's rays throughout the year. An antifreeze liquid circulates through the solar collectors, absorbs heat, and then is pumped through a heat exchanger. The heat exchanger transfers the heat to a water-filled, insulated storage tank. The water in this tank is then circulated through a fan-coil unit which heats the air that is circulated through the house.

FROM SUNLIGHT TO ELECTRICITY

Another major step forward in the development of a solar energy industry seeks to collect sunlight, focus the light from a large collecting area onto a small area, convert the solar energy to heat, use the heat to produce steam, and then use the steam to generate electricity. Pilot projects are underway in which large fields of mirrors or lenses focus the sunlight that falls on one thousand square units of surface onto a single unit of surface. This means that there will be nearly one thousand times as much energy to heat the small surface.

A large solar furnace has been built in the French Pyrenees mountains that collects and focuses enough sunlight to create temperatures almost ten times as great as can be produced by a Bunsen burner. Obviously, a large collecting surface designed so as to reflect or focus the light onto a small area can produce temperatures high enough to change water to steam.

Another type of solar thermal power system collects and concentrates sunlight and converts the light to heat in a series of mirrors. A fluid circulates among the collecting mirrors and transfers the heat to a central boiler or steam turbine where electricity is generated.

A five thousand-watt solar thermal power experimental system has been built in Albuquerque, New Mexico, by the U.S. Department of Energy. This small-scale system is being used to determine the most efficient ways to collect solar energy and convert it to electricity.

Job and career opportunities in this branch of solar energy will very probably develop first in the southwestern United States. These opportunities will include not only the design, construction, and operation of solar thermal power stations but also the development of ways to make sure the stations have minimum harmful impact on the environment.

Direct Conversion

The story of direct conversion of solar energy to electricity is told in a publication of the Mobil Tyco Solar Energy Corporation. The Mobil Oil Company has been active in developing this part of the solar energy industry. The following quotation is from a speech by Richard F. Tucker, former chairman of Mobil and president of Mobil Oil Corporation.

> "The direct conversion of sunlight into electricity is made possible by the photovoltaic process, in which an electromotive force—voltage—is generated as a result of ionizing radiation—sunlight,
>
> Photovoltaics have been around more than 70 years. But there wasn't a great deal of interest in the process until the mid-50s, when the first silicon solar cell was made. Earlier cells

made of other materials were extremely inefficient in terms of their ability to convert sunlight into electricity. The silicon cell made it possible for the first time to effectively convert sunlight into electricity.

As with so many other technical achievements, the United States space program gave impetus to development of the silicon solar cell. The equipment on board more than 600 manned and unmanned spacecraft, including the massive Skylab, was powered by solar cells.

But the NASA program was a "money-is-no-object" effort; the cells cost between $100,000 and $200,000 per kilowatt of power. Units designed for use on earth run about $15,000 per kilowatt. Our goal is to reduce these costs to $500 per kilowatt. This is the level at which photovoltaics would be able to compete with centrally generated forms.

The principal reason why silicon solar cells cost so much is the way they are made. The traditional method is to start with high-purity silicon, melt it at 2650°F and 'grow' a sausage-shaped ingot about three inches in diameter. Once the ingots have been 'grown,' they are sliced and polished into twelve-thousandths-of-an-inch-thick wafers. This process is slow and extremely wasteful.

But Mobil Tyco has a better idea. It's called edge-defined, film-fed growth (EFG), and we think it holds promise for ultimately bringing solar electricity down to a level comparable to those for conventional generating techniques.

With the EFG process, the molten silicon is drawn upward by capillary action through a die mounted in the crucible. As it emerges from the die, the liquid crystallizes, and this continuously forming crystal is pulled slowly upward. The result is a continuous ribbon of virtually single crystal silicon, one inch wide and about eight-thousandths of an inch-thick . . . By growing the ribbon into the dimensions required for the cell, there is virtually no scrap, and the costly slicing and polishing stages are eliminated.

A typical solar cell contains two layers of silicon with an outside wire attached. In one layer, a few atoms in the silicon crystal have been replaced by boron atoms; in the other, the replacement atoms are phosphorus. Sunlight falling on the cell drives electrons along the wire from the phosphorus-silicon layer to the boron-silicon layer and the electrons can be made to do the things electricity does as they move through the wire.

Solar cells are wired together to form solar modules and these modules form the building blocks of solar electric systems. A module of forty cells, for example, will provide enough electricity to charge a twelve-volt automobile battery. But it would take a twenty-by-thirty-foot panel of solar cells, operating at 10 percent efficiency and with a peak output of five thousand watts at midday in the northeastern United States, to yield enough electrical power for an average home.

Improving the efficiency of all phases of solar energy development is a big part of employment in this industry. Recently, thin-film solar cells solved several problems. In these cells, a much less expensive form of silicon is deposited on plastic film. They can be mass-produced and are more efficient than conventional cells.

According to an information bulletin published by the U.S. Department of Energy, "The most common applications for photovoltaics (PV) are remote or *stand-alone* systems." Such systems are particularly advantageous in remote locations where there are no power lines. Worldwide, about twenty thousand homes rely on this source of electricity. PV is also used for powering appliances in motor homes, and in the less technologically developed nations, it is used to generate electricity for refrigerating medicine, pumping water, or lighting villages.

The U.S. Coast Guard operates many PV systems in buoys and shoreline markers. By backing up PV systems with generators powered by diesel engines, wind power, or hydroelectric generators, solar power is finding new uses where the possibility of too little sunshine is a problem.

Solar Power Satellite

The challenge of making the sun's energy available as a replacement for other finite supplies appeals to a wide variety of people—

from backyard mechanics to highly trained engineers and scientists. Peter Glaser represents those who combine training, experience, and unbridled imagination. When he proclaims the positive features of his solar power satellite solution to humanity's energy supply problems, his less imaginative colleagues wonder if he is confusing reality with science fiction. But rarely are patents issued for inventions that fail to measure up to hard-nosed reality. Quoting from U.S. Patent No. 3,781,647, Glaser's invention is intended "to provide an integrated system capable of providing electrical power on earth in large amounts from solar energy" and "is arranged for a continual delivery of electrical power free from such factors as cloud cover, eclipses of the sun by the earth, and the like." Furthermore, "It is yet another primary object of this invention to provide electrical energy in large amounts by a system and method which do not deplete the earth's natural resources and which are essentially pollution free."

By this invention, "the radiation energy derived from the sun is converted to microwave energy in equipment maintained in outer space, then it is transmitted as microwave energy to suitable collectors on earth." As seen by Glaser, this system minimizes the "major drawbacks associated with the direct terrestrial collection of solar energy" because "the problems of absorption of the solar radiation by the atmosphere are eliminated" and "the microwave energy can be collected in locations on earth without regard to availability of solar radiation."

The size of the satellite collector and transmission systems invite the science fiction allegation. To meet the power requirements of the northeastern United States, for example, would require a collector area of approximately nine square miles. This is equivalent to a circular area more than three miles in diameter. Such a collector would weigh more than 330,000 pounds. Glaser is well aware that the future of the space power satellite system depends upon the development of launching facilities that are capable of placing enormous payloads into low-earth orbits and a space transportation system, including launch vehicles, of much greater capacity and efficiency than those already on hand. However, 95 percent of the required construction material for an SPS can be obtained from lunar resources, thus greatly reducing launch requirements. He knows there are technical, economic, environmental, and societal issues that are scaring away many of the people whose decisions control the future of solar satellite power transmission. But he believes that, "thus far, studies by European, Japanese, U.S., and Soviet teams have shown that there are no likely show stoppers in an SPS program."

Obviously, the idea of a career involving a space power station appeals only to people who are able to let their imaginations soar a bit. But modern research is rapidly changing the whole field of space exploration. A few years ago it took one hundred pounds of instruments to harvest one kilowatt of solar energy; today, less than fourteen pounds. Present-day solar energy collecting cells are about 10 percent efficient. It may be that tomorrow's cells will be mounted on very lightweight, highly efficient laminated plastic blankets. The whole idea of space power stations assumes that engineering research will move rapidly toward solving the problems that must be solved before people can commute between the earth and space platforms. But this is the kind of idea that is sure to be appealing to at least a few people who are thinking about future careers.

Recalling the theme of the opening paragraphs of this chapter, solar energy holds promise of heading off the serious problems that our nation would face if our demands for energy exceed the earth-bound supply. If the "good news" aspects of solar energy are to overcome the "bad news" aspects, people will be needed whose training and motivations enable them to solve very difficult problems—problems arising from the very dilute form in which solar energy arrives together with the storage problems that must be solved before solar energy can be distributed night and day, year in and year out.

SOLAR ENERGY JOBS AT A GLANCE

Engineers and Scientists
(Generally requires a four-year bachelor's degree and may require additional postgraduate study)

solar engineers architects
electrical engineers environmental engineers
design engineers civil engineers

Support Workers
(Generally requires at least a two-year associate degree for technicians; technologists require a four-year bachelor's degree)

architectural technicians engineering technologists

Blue Collar Workers
(Generally requires apprenticeship)

sheet metal workers ironworkers
plumbers welders
electricians

Semiskilled Workers
(May require high school education)

laborers equipment operators

CHAPTER 7

CAREERS IN THE NUCLEAR ENERGY INDUSTRY

Certain inescapable facts suggest that the number and attractiveness of career opportunities in the nuclear energy industry will increase, particularly after 2010. This prediction is based on an energy policy statement presented to the American Congress by the Executive Department in 1991. Whereas nuclear energy was used to generate 11 percent of the nation's electricity in 1990, it is predicted that this will decrease to 10 percent by 2010 but will increase to 16 percent by 2030. Closely related is the prediction that the use of oil and gas to produce electricity will increase from 7 percent to 10 percent but will then decrease to 4 percent during these same time intervals.

Few other energy-related career choices confront young people with more difficult decisions. Powerful interests oppose the development and utilization of nuclear energy. This opposition stems from (1) concern about how safely nuclear reactors can be operated, (2) concern about the safe disposal of radioactive wastes, (3) lack of faith in governmental licensing and regulatory action, and (4) uneasiness lest the expansion of peaceful uses of nuclear energy foster its use in weaponry. The best we can do here is to try to put the nuclear energy industry into a perspective that is consistent with the sociopolitical and economic interests of those who oppose as well as those who promote the development and utilization of nuclear energy.

HISTORY OF THE NUCLEAR ENERGY INDUSTRY

Humanity's ability to manage nuclear energy began in that primeval urge to understand our surroundings and ourselves. Nuclear energy comes from within atoms, and atoms are nature's building blocks. Appreciating all of nature's mysteries and taking advantage of everything nature has to offer begins with understanding how energy enables substances to be taken apart and their building blocks reassembled into substances with new and more valuable properties.

The knowledge that culminated in releasing and controlling the energy within atoms began shaping up far back along the stream of civilization. Of interest here are several lines of investigation that reach back into the 1800s. Early in the 1900s, a series of discoveries proved that all of the elements are made up from a few kinds of subatomic particles. Chief among these particles are protons, electrons, and neutrons. In general, the only differences in the atoms of iron and gold, arsenic and oxygen, and in all other elements are due to differences in the numbers of subatomic particles that are used to build their atoms.

Atoms of uranium, the heaviest of nature's elements, contain the greatest number of these subatomic particles. Each uranium atom contains 92 protons and 92 electrons. The predominant variety or isotope of uranium contains 146 neutrons but more about that later. It was but a matter of time until someone would come up with the idea that totally new elements might be created by forcing additional subatomic particles to lodge in the nuclei of uranium atoms.

Albert Einstein

Closely related to this idea was the equation that Albert Einstein shaped up and announced in 1905. The simple expression, $E = mc^2$, enabled Einstein to share with his colleagues in science his theory that energy (E) and matter (m) are interchangeable. It is the c^2 part of the equation that accounts for the enormous dimensions of this interchange. The c represents the speed of light, a speed so great (186,282 miles per second) that it boggles the mind even without multiplying it by itself. For example, to change a mass no greater than that of a sheet of typing paper into energy would yield a quantity of energy equal to thousands of times the amount of electrical energy consumed by an average household each month.

Enrico Fermi

All of an atom's protons and neutrons are packed into an amazingly dense nucleus; the atom's electrons are distributed around the nucleus. It was believed by several scientists, particularly Enrico Fermi, that it might be possible to force neutrons to lodge in the nuclei of uranium atoms. Once one or more new neutrons had been lodged in a uranium nucleus, it was believed that a neutron might split into a proton and an electron. By adding one more proton to the nucleus of a uranium atom, a new kind of element would be formed.

Fermi designed an experiment that would allow him to prove or disprove this hypothesis. But the results of his experiments kept coming out differently than he had expected. Rather than creating new atoms only slightly more massive than uranium, he found atoms appearing in his equipment with roughly half the mass of uranium atoms—barium and krypton, for example.

Fermi published the results of his experiments in the 1930s, including his curiosity about the appearance of barium and krypton atoms in his apparatus. Among the scientists who were puzzled by Fermi's results were Lise Meitner and her colleague, Otto Frisch, They were first to publish the possibility that the uranium atoms were actually being split into roughly equal fragments. Within months, other scientists calculated the amount of energy that would have to be released if this was really what happened and performed additional experiments to confirm the hypothesis.

Much courage was needed to say that bombarding neutrons had split or fissioned the uranium atoms. The equation $E = mc^2$ figures in this courage. The two atomic fragments simply did not weigh as much as the uranium atom that was split. When 92 protons and 146 neutrons are packed together to make a uranium nucleus, a certain amount of mass is converted to energy. No uranium atom could split into a barium and krypton atom without additional mass being converted to energy—quite a lot of energy.

Another part of this story played a vital role in the ultimate release and control of nuclear energy. When the uranium nucleus splits into roughly equal halves after being hit by a bombarding neutron, two or more neutrons are ejected when the uranium nucleus splits. Each of these neutrons stands a chance of lodging in the nucleus of a nearby uranium atom and setting off another fission event in chain fashion. On this basis, although the energy released when the first uranium atom is split is less than the amount of energy needed to create a bombarding neutron, the continuing chain reaction not only "pays back" this energy but begins to yield a net energy gain. Since the total chain reaction occurs in a fraction of a second, the energy that is released by the total mass of uranium atoms becomes the enormous quantity of energy associated with nuclear fission reactions.

Technology developed ways to control the chances that neutrons being ejected from splitting uranium atoms would actually lodge in nearby atoms and continue the chain reaction. Thus, the rate of the chain reaction could be controlled and with this control came the release of nuclear energy for peacetime purposes.

Concentrating Uranium

Uranium-fueled reactors require uranium that is highly refined. But an even more difficult process must be carried out before uranium can be used for reactor fuel. Uranium as it is found in nature consists of two kinds of atoms or isotopes. One kind of uranium atom has three fewer neutrons per nucleus than the other. In other words, although all uranium atoms must contain exactly 92 protons, uranium atoms may have either 143 or 146 neutrons per atom. The two kinds of atoms are called the uranium-235 and uranium-238 isotopes.

Only the uranium-235 isotope will serve as reactor fuel. But fewer than one of every hundred atoms in a batch of newly mined uranium is of the uranium-235 variety. Reactor fuel must consist of approximately three times this many uranium-235 atoms. It is no easy problem to sort out and concentrate the uranium-235 atoms in a batch of uranium.

Learning to concentrate the uranium-235 atoms in samples of uranium was a major step in making nuclear energy available. Since all uranium atoms have the same number of protons and electrons and since the chemical properties of atoms depend on the numbers of electrons in their make-up, no chemical process could be used to separate the two isotopes of uranium. In other words, no chemical could be found that would precipitate one kind of uranium isotope from a solution that would not at the same time precipitate the other.

But the two isotopes do have different masses. The uranium atoms with three more neutrons per nucleus are heavier. If uranium can be combined with another element to form a gaseous compound, the molecules of the compound that contains the heavier-mass uranium isotope will be heavier than those molecules containing the lighter uranium isotope. It is known that gas molecules diffuse or spread about at rates that depend upon the mass of the molecules.

These facts were used to come up with a process for concentrating the uranium-235 atoms in a sample of uranium. The uranium was made into uranium fluoride, a gaseous compound. The gas was then forced to diffuse through a long series of porous barriers. The heavier uranium-238 atoms could not find their way through the barriers as easily as the lighter uranium-235 atoms. In time, the uranium atoms arriving at the end of the diffusion "obstacle course" contained a higher percentage of the lighter uranium-235 atoms. The uranium was then recovered from the gaseous compound and converted to a metal ready to be fashioned into fuel for use in reactors.

Today there are also centripetal and spectroscopic methods of separating uranium-238 and uranium-235.

TRENDS IN EMPLOYMENT

Approximately 9,500 people are needed to prospect for uranium ore and to operate uranium mines. Another 1,700 people are employed in extracting the uranium from the ore. Some uranium mines are underground, others are open-pit mines. Jobs in these mines are pretty much like those in coal mines and are located mainly in New Mexico, Wyoming, Utah, Colorado, and Arizona. Uranium ore processing mills provide jobs for chemical process operators, electricians, and engineers. Machinery installers and repairers, pipe fitters, and carpenters are also needed.

The data in Table 7.1 bear on the number of employment opportunities in the nuclear energy field, both today and in the near future. The data are also valuable because they show the wide variety of jobs that exist in this field.

Many factors make the prediction of job opportunities in the nuclear field very hazardous. Much depends upon the number of new nuclear power stations that will go "on stream." There are 82 nuclear plants in operation across the country, but according to *Time* magazine (February 13, 1984) no utility company has placed an order for a new nuclear plant since 1978. "The nuclear industry," *Time* reported . . . "is not well. . . . It has been suffering seriously for nearly five years."

A more recent series of setbacks has made the condition of the industry worse. In July 1983, the Washington Public Power Supply System, a consortium of 23 electric companies which, having cancelled or postponed construction of four of its five projected nuclear power plants, defaulted on $2.25 billion in bonds. In January 1984 the Nuclear Regulatory Commission refused to grant Commonwealth Edison a license to operate an almost completed $3.7 billion plant at Byron, Illinois, because of dubious quality-control construction procedures. Work on a Marble Hill, Indiana, nuclear plant (half built at a cost of $2.5 billion) was canceled by the Public Service Company of Indiana, and work on the Zimmer plant at Moscow, Ohio, was halted; the companies involved had decided that it would be converted to a less expensive coal-burning facility. It is not improbable

Table 7.1: Nuclear-Related Employment by Occupation, 1983, 1985, 1987, 1989 *(continued)*

	1983	1985	1987	1989
TOTAL EMPLOYMENT	282,800	279,400	285,800	291,400
Managers	23,850	26,500	28,200	28,500
Engineers	57,600	57,950	59,850	55,050
Chemical engineers	2,800	2,650	3,000	2,900
Civil engineers	4,950	4,650	4,550	3,600
Electrical & electronics engineers	8,900	8,150	9,750	9,250
Mechanical engineers	14,050	12,100	14,300	12,750
Nuclear & reactor engineers	8,850	9,250	9,900	10,350
Metallurgical engineers	900	950	850	1,050
All other engineers	17,150	20,200	17,500	15,150
Mathematicians	1,750	1,900	2,250	2,200
Physical Scientists	9,500	9,250	9,900	9,500
Chemists	3,100	2,850	2,900	2,900
Geologists & geophysicists	500	500	500	650
Physicists	3,700	4,000	3,950	3,650
Metallurgists	500	400	400	350
Other physical scientists	1,700	1,500	2,150	1,950
Life Scientists	4,650	4,450	4,750	5,300
Biological scientists	1,500	1,200	1,050	1,000
Health physicists	2,700	2,900	3,050	3,350
Other life scientists	450	350	650	950
Other Professional Workers	22,850	25,700	26,450	27,950
Technicians	57,600	49,150	55,600	54,900
Drafters	10,300	6,300	4,800	4,400
Electrical technicians	8,550	8,150	8,850	9,200
Other engineering technicians	7,400	6,600	7,800	6,900
Physical science technicians	3,800	3,050	4,050	4,050
Life science technicians	1,000	700	650	800

(continued)

Table 7.1: Nuclear-Related Employment by Occupation, 1983, 1985, 1987, 1989 (continued)

	1983	1985	1987	1989
Technicians *(continued)*				
Health physics technicians	5,900	5,950	7,850	8,450
Senior nuclear reactor operators	1,400	1,800	2,250	2,150
Nuclear reactor operators	1,750	1,850	2,000	2,000
Auxiliary reactor operators	5,300	4,950	5,000	4,600
All other technicians	12,200	9,800	12,350	12,350
Skilled Craft Workers	**34,000**	**37,200**	**32,700**	**37,550**
Welders	2,300	2,600	3,150	2,250
Other skilled craft workers	31,700	34,600	29,550	35,300
Clerical Workers	**26,800**	**23,600**	**20,750**	**21,000**
All Other Workers	**44,200**	**43,700**	**45,350**	**49,450**

NOTE: Revisions were made to previous data in 1989.

Source: U.S. Department of Energy, Office of Energy Research.

that some of the forty-eight plants presently under construction may also be canceled.

Behind these failures lies a variety of causes, including problems created by greatly underestimating the cost of constructing a nuclear facility (with cost overruns of from twice to fifteen times the original projections) and the speed with which it could be done; problems of poor supervision along with faulty or inadequate construction, which have resulted in a series of potentially serious accidents; and the problem created by the growth of a strong national antinuclear movement.

The antinuclear movement is a coalition of diverse groups whose ranks include well-established and new environmental groups, doctors, movie stars, physicists, and other concerned individuals and organizations. In its battle against nuclear energy, by using such tactics as blockading nuclear sites and engaging in time-consuming legal procedures, this coalition has succeeded in disrupting and delaying construction and operation of a number of nuclear reactors.

In spite of these difficult problems, many people involved in the energy field, from utilities executives to nuclear physicists to government officials, believe that the widespread use of nuclear energy is inevitable. They believe that if we are to satisfy U.S. power needs and keep the economy healthy we must have a *combination* of energy sources, and they see many more nuclear plants in our country's future.

The future of nuclear energy development is also tied to public acceptance of the so-called breeder reactor. Some energy futurists warn against burning up the nation's supply of uranium in ordinary fission reactors. They argue that the uranium on hand should be stockpiled and that no more ore should be harvested until breeder reactors are ready to go on stream.

Breeder reactors create more nuclear fuel than they consume. This statement appears to contradict the law of conservation of energy. The explanation follows:

The secret of the breeder reactor lies in using the scarce uranium-235 to produce, or breed, other kinds of nuclear fuel. The process is simple in theory. Build an ordinary nuclear fission reactor. Fuel it with the usual uranium-235 enriched fuel. But pack the reaction chamber with the more abundant uranium-238. When the uranium-235 atoms undergo fission, in addition to the fission fragments that are produced, high speed neutrons are hurled from the nuclei of the uranium-235 atoms.

Some of these neutrons will lodge in the nuclei of the surrounding uranium-238 atoms and things will begin to happen. Before the uranium-238 atoms were hit by the high-speed neutrons, their nuclei consisted of 92 protons and 146 neutrons—a reasonably stable, nonfissionable arrangement of protons and neutrons. But when one additional neutron enters such a nucleus, two neutrons lose beta particles—that is, tiny bundles of negative electricity. The loss of these two beta particles changes two neutrons to two protons.

By changing the number of protons in the nuclei of these two atoms, they now become plutonium-239 atoms. Plutonium-239 is a fissionable nuclear fuel. In practice, a greater amount of plutonium-239 is produced than the amount of uranium-235 that was consumed. It is estimated, for example, that if all of the uranium in the United States were to be used only for breeder reactors, the electricity that could be generated would meet current consumption for sixty-four thousand years.

THE SAFETY AND SANITY OF NUCLEAR REACTORS

How safe are nuclear reactors, especially the breeder type? This question is especially important to people who are contemplating careers in the nuclear energy field. For people who are concerned about international relations, there is an additional question. Does the availability of more and more fissionable nuclear material in the world pose additional threats to world peace?

Answers to these questions depend upon whom you ask. Both questions run up against deep-seated points of view, prejudices, beliefs, and traditions. One reaction is to dodge the basic issues and raise counteracting questions such as: How safe is any large-scale source of energy? Aren't people killed or injured every day in accidents around energy conversion systems? Is it more of a threat to world peace to stockpile fissionable material that could be used to make bombs than to have one nation become dependent upon another for its fuel supply?

The questions raised in these arguments point to a significant group of job and career opportunities indirectly related to the design, building, and operating of nuclear power stations. People are needed who can inform the public about the nature of nuclear energy and about the pros and cons of its development. The social and political sciences are as much involved in these areas as are the physical sciences.

Controlling Nuclear Reactions

Basic to the theory of nuclear fission reactors is bringing together a large enough quantity of uranium-235 or other fissionable material to cause a chain reaction. In any quantity of uranium-235, a known fraction of its atoms will spontaneously eject neutrons each moment. Each neutron that is ejected has enough energy to lodge in the nucleus of a neighboring uranium-235 atom. Whether the neutron does or does not hit another atom depends upon how many atoms are in the neighborhood.

In nuclear reactors, less than "critical mass" amounts of the nuclear fuel are packed in long, carefully sealed metal tubes. The reaction chamber is built in such a way that forty thousand or more of these tubes can be packed or bundled into fuel units. The fuel units are arranged in the reactor vessel in a manner that allows fluid coolant to flow between them and remove the heat that is generated by the fission process.

Interspersed among the fuel rods are control rods made out of material that absorbs or captures neutrons. The idea is that if the neutrons that are ejected by one fissioning uranium atom are "soaked up" before they hit another uranium atom, the chain reaction can be controlled. This allows only a carefully calculated number of uranium atoms to be fissioning at any one time. When more heat from the reactor is called for, the control rods are pulled out of the core and the chain reaction is permitted to involve a greater number of uranium atoms.

So long as the fuel in the rods is kept from clumping together enough to form a critical mass, there is no risk of the reactor becoming an uncontrolled atomic bomb. Fail-Safe engineering is intended to insure that any accident or human or mechanical error that might allow too many fuel rods to get too close to each other automatically triggers machinery that forces the control rods into a safety position and the reactor is shut down. This procedure should head off the possibility of the fission reaction occurring. But, as in the 1975 Browns Ferry, Alabama, and the Three-Mile Island, Pennsylvania, accidents, human and mechanical errors can combine to increase the possibility of a disastrous meltdown.

Chernobyl and Other Lessons

On April 26, 1986, an event occurred at Chernobyl in the Soviet Union that has had much to do with how people feel about the development and utilization of nuclear energy. On that date, an experiment being conducted in a reactor's electrical generator was apparently not adequately monitored. The reactor produced excessive energy and the water cooling system could not handle the resulting heat. When the water changed to steam, it could no longer absorb the excessive neutrons and thereby bring the reactor under control. In turn, the reactor produced more energy, and more of the cooling water changed to steam. Fewer neutrons were absorbed by water, and more of them were permitted to fission additional uranium-235 atoms. The released energy increased in chain fashion until the reactor became totally out of control. The resulting immediate release of energy melted the reactor into the equivalent of a small-scale atomic weapon with subsequent catastrophic loss of thirty lives and the conversion of the surrounding countryside into a veritable wasteland. More deaths from radiation followed, and it will be years before the long-term health effects are fully understood.

It has been reported that the Chernobyl reactor was under the management of an electrical engineer who was inadequately trained, and the operating crew on duty at the time was overtired and under pressure to complete their assigned experiment. Mandatory safety rules were disregarded. Furthermore, because of an increasing need for electric power in the Soviet Union, the concentration of uranium-235 in this reactor had been increased to the point where automatic safety measures were no longer adequate.

Lessons learned at Chernobyl were terribly expensive, but their messages may well point the way toward insuring that nuclear energy can play a positive role in providing adequate supplies of energy far into the future. For sure, these lessons have much to say to people who accept the challenges that come with careers in the nuclear energy industry.

In keeping with the intent to put into perspective the prevailing attitudes toward careers in the nuclear energy industry, several things need to be emphasized. Prior to 1939, the motivations underlying the study of nuclear energy were primarily those that stem from the urge to understand nature's events and circumstances. What was being done in scientists' laboratories went on pretty much apart from the events in the sociopolitical world. But it is not to be expected that scientists would be unaware of or insensitive to society's problems.

The much-cited 1939 letter from Albert Einstein to President Franklin D. Roosevelt provides a persuasive case in point. In effect, this letter said, "It may be possible to build extremely powerful bombs which, although too heavy for air transport, could be carried by boat and exploded in a port and might very well destroy the whole port together with some of the surrounding territory." Einstein was aware of and sensitive to the threats to humanity's well-being that were developing in Europe, and acted accordingly.

Equally significant, U.S. scientists responded to the threats posed by Japan's actions that began on December 7, 1941, at Pearl Harbor. Aided by a burgeoning technology, their response culminated in the United States dropping two atomic bombs on Japan in 1945. The deaths and destruction suffered at Hiroshima and Nagasaki, and the radiation effects on survivors, are not easily forgotten.

Chernobyl, Hiroshima and Nagasaki, Brown's Ferry and Three-Mile Island, bomb tests, and the awesome threat of nuclear weaponry will always loom large whenever one considers a career in the nuclear energy industry. At the same time, nuclear energy stands to continue to play a major role in humanity's affairs. Few other careers require us to keep in mind the circumstances that have been responsible for how people feel about this energy resource and to make the kinds of decisions that will be best for oneself and for society.

Radioactivity

One hazard associated with nuclear reactors worries many people. Whenever things happen to the nuclei of atoms there is the possibility that atomic fragments or bits of energy will be hurled from the atoms. This is called radioactivity, and there is certainly much radioactivity associated with nuclear reactors. The radioactivity that is associated with reactors, however, is taken into account by the engineers who design them and the required shielding is built into the reactors to protect the people who must work in their vicinity.

Fission by-products or waste materials from reactors create a more involved problem. In the chaotic conditions accompanying the splitting of the uranium atoms in the atomic fuel, only rarely are the fission products stable atoms. Nearly all are radioactive. Some of these reactor waste products disintegrate in a matter of seconds and become stable, nonradioactive atoms. Others wait for many years to complete the rearrangement of their nuclei and become stable atoms. Spent fuel rods, contaminated with various fission products including plutonium-239, will remain radioactive for many thousands of years.

Radioactive forms of the elements cannot be allowed to become scattered throughout the environment. If radioactive calcium were scattered throughout the countryside, for example, it could be picked up by growing plants. The plants, if eaten by a cow, might pass the radioactive calcium along so that it would show up in the cow's milk. People who drank the milk would take the radioactive calcium into their bodies.

Theoretically, each radioactive calcium atom in a person's body could be a threat to the health and well-being of the person. Suppose, for example, that a radioactive calcium atom is built into a complex molecule that plays a vital role in one of the life processes. Before the molecule has a chance to do what it ordinarily does, the radioactive calcium atom might eject a beta particle.

In effect, this adds another proton to the nucleus of the atom and the atom is no longer calcium. It is now a scandium atom. Scandium atoms do not have the same chemical properties as calcium atoms. Consequently, the new scandium atom cannot fit into the original complex molecule. This can destroy the whole molecule or at least disturb the proper role of the molecule in a life process.

Radioactive atoms taken up in body tissues may pose another kind of threat to our well-being. The particles or bits of energy that are ejected when radioactive atoms disintegrate are a threat to the "glue" that holds all kinds of molecules together. A beta particle whizzing through a nearby molecule is very much like an electron. Atoms are held together to form molecules by sharing electrons. An electron "on the loose" can easily break the bond between two atoms in a molecule and the molecule would come apart.

Suppose a radioactive atom lodges in body tissue near a person's reproductive organs. If this atom undergoes a rearrangement of its nuclear particles and ejects a beta particle, undesirable things can happen. Unless the beta particle spins off its energy by hitting less significant atoms, it might penetrate tissues and hit a molecule that is a part of a gene in one of the chromosomes of an egg or a sperm cell. The atoms making up this molecule could be rearranged. The molecule, in turn, would take on new properties and whatever trait is controlled by that gene could be changed—that is, undergo a mutation.

If so, there is the chance that the characteristics of a new baby would be influenced by the radioactivity of the original unstable atom that ejected the beta particle. The odds against this sort of thing happening are almost astronomical. But it can happen. People who are trained to work around radioactive materials use special clothing, procedures, and decontamination chambers to protect themselves to increase the odds against such an event happening to them.

People who work with radioactive materials know that all body tissues are exposed constantly to radioactivity from natural sources. Many of the elements that are essential for body processes occur in radioactive forms. Cosmic rays continually expose body tissues to high energy impacts that can create effects very similar to those caused by radiation from unstable atoms. Part of their job is to understand radiation, and to prepare for handling it safely.

There are many jobs for people who specialize in protecting reactor personnel from the harmful effects from radiation. These people know exactly how far each kind of radioactive atom can eject its fragments and how much shielding material is needed to absorb its energy safely.

NUCLEAR ENERGY JOBS AT A GLANCE

Engineers and Scientists
(Generally requires a four-year bachelor's degree and often requires postgraduate study)

nuclear engineers	electrical engineers
radiation chemists	mathematicians
nuclear physicists	radiological physicists
health physicists	radiological chemists

Support Workers
(Generally requires at least a two-year associate degree and may require a four-year bachelor's degree)

radiation monitors	hot-cell technicians
nuclear reaction operators	decontamination workers
accelerator operators	waste-treatment operators
radiographers	waste disposal workers
radioisotope-production operators	quality control technician

Blue Collar Workers
(Generally requires apprenticeship)

pipe fitters	sheet metal workers
electricians	boilermakers
carpenters	bricklayers
welders	painters

Semiskilled Workers
(May require high school education)

maintenance workers	messengers
laborers	

CHAPTER 8

CAREERS IN THE ELECTRIC POWER INDUSTRY

It takes only a brief blackout or interruption of electrical power to make us realize how utterly dependent we are on the electric power industry. When a storm or some other emergency knocks out the system of wires and transformers which brings electricity into our homes, we begin groping almost immediately for a source of light, worrying about our home becoming too cold or too hot before the heating or air-conditioning system comes back on, wondering how we can prepare the next meal, worrying about how long the frozen foods in the deep-freeze will keep, and fussing about similar threats to our comfort and well-being.

The 550,000 men and women who hold jobs in the electric power industry provide a vital service to the American people. To be aware of how important it is to generate, transmit, and distribute adequate electrical power night and day, year in and year out, to nearly every urban home and to 98 percent of the nation's rural homes may well motivate public-service-minded people to seek jobs and careers in the electric power industry.

To deliver electricity to all customers whenever it is needed and in quantities adequate to meet the demand is the challenge to be met by an electric power company. An electric utility system uses generators to transform waterpower or the energy in coal, oil, gas, or uranium into electrical energy. Systems of transformers in switchyards and on local light poles adjust the voltage so that the electrical energy can be sent efficiently over long distances and be brought to the consumer safely.

Widely different kinds of job and career opportunities exist in the electric power industry. Some 200,000 people are employed in jobs involving the generation, transmission, and distribution of electricity. An even larger number of people are employed in engineering, scientific, administrative, sales, clerical, and maintenance jobs.

HISTORY OF THE ELECTRIC POWER INDUSTRY

Strictly speaking, electricity is not an energy "resource" in the sense that there is somewhere in the environment a supply of electrical energy waiting to be tapped and distributed. Actually, electricity is simply a highly convenient form of energy that is obtained by exploiting other energy resources, especially the fossil fuels, waterpower, and more recently, atomic energy.

The electrical industry's role in making energy available whenever we need it and in whatever quantities we can afford to pay for is to convert energy from one or another source to electrical energy and to transport this electrical energy to us.

Thomas Edison

The early history of the electric power industry is touched on by Jack Egan's story of the roots of the General Electric Corporation which appeared in *The Washington Post* on October 1, 1978. He traced the origin of GE to October 15, 1878, when a group of investors raised $50,000 to back Thomas A. Edison's search for a successful electric light bulb. This led to the organization of the Edison Electric Light Company, with Edison—who was then thirty-one years old—holding half of the company's stock.

Within a year, Edison announced that his use of a carbonized sewing thread for the filament produced a light bulb that burned for forty hours. Perhaps it was the inventor's successful record with other inventions that caused people to believe that a truly efficient light bulb was on the way, since his announcement on December 21, 1879, created an effect on the stock market. Edison Electric Light Company stock rose sharply, eventually hitting $3,500 a share while gas company stocks fell in price.

The promise of efficient electric light bulbs was what it took to trigger rapid development of the electric power industry. Quoting Egan, "The first full-scale introduction of an electrical system came in London. But the showpiece was the unveiling of the Pearl Street Station in New York City in 1882 which eventually served 946 customers and 14,311 incandescent lamps."

Egan quotes Edison's interview by the *New York Times:* "The giant dynamos were started up at three in the afternoon and, according to Mr. Edison, they will go forever unless stopped by an earthquake." As Egan pointed out, it was easier for Edison to anticipate the hazard of an earthquake than to foresee the threat of a lightning storm or whatever it was that caused New York City's famous blackout some eighty years later.

The first electric power company in the United States was the California Electric Light Company in San Francisco. In 1879, this company installed two dynamos of the type invented two years earlier by Charles Francis Brush. At the flat rate of $10 per lamp per week, the company could deliver enough electricity to keep twenty-two lamps burning.

Michael Faraday

All dynamos or electric generators apply a principle discovered primarily by Michael Faraday between 1821 and 1831. First, Faraday discovered that a magnetic field is set up around a wire that is carrying an electric current and then he discovered that an electric current is induced in a wire that is made to move through a magnetic field. Putting these ideas together, Faraday realized that large quantities of electrical energy could be generated if large coils of wire are made to rotate through magnetic fields set up between the poles of strong magnets. And the best way to create strong magnets is to send strong electrical currents through large coils of wire around iron cores.

Steam engines, windmills, or waterpower could be used to force the coils of wire to rotate between the poles of the magnets. Some of the electricity that was generated could be turned back to increase the strength of the magnetic field between the poles. But don't forget the well-known law of energy conservation. The electricity that was generated produced a magnetic field that was opposed by the magnetic field that was generated around the coils of rotating wire. By overcoming this opposing force with the energy produced by a steam engine or waterwheel, electrical energy was exchanged for the energy originally present in the steam that drove the steam engine or waterwheel.

Electricity's Versatility

The history of the electric power industry since its beginnings has consisted pretty much of finding sources of energy to convert to electrical energy and to do the conversion as efficiently and as economically as possible. The convenience of energy in the form of electricity almost creates markets automatically. The heating effects produced when electricity flows through high resistance conductors are conveniently taken advantage of in many kinds of cooking and heating devices. The electromagnetic effects are applied in the design of telegraph and telephone systems and in more modern inventions that use electromagnetic waves such as television, radio, radar, and similar telecommunication systems.

By feeding electricity into a slightly modified dynamo or generator, the dynamo or generator becomes an electric motor. And electric motors do all manner of chores in today's homes and industries. The electronic effects of electricity find almost as wide an array of applications, ranging from doing our arithmetic for us in pocket calculators to translating the sound track on movie film into music and dialogue.

APPLYING FOR A JOB

We have emphasized how important it is to know as much as possible about the specific energy resource that supports the industry or organization in which you seek a job or career. We have also emphasized the importance of being aware of what energy is, how it is obtained, and how it is transported.

Using the electric power industry as an example, we now turn to the specific things a person must do to obtain a job. The logic is that the application blanks that must be filled in, the interviews that are arranged, and the general procedures for getting a job are similar in many energy-related employment situations.

SOME JOB DESCRIPTIONS

The following examples of job descriptions or specifications are interesting because they show exactly the kinds of things employees in the electric power industry are hired to do.

Performance Engineer

Function: The performance engineer is responsible for supervising and conducting tests and inspections to calculate and determine the efficiency of power plant equipment; performing work involved in supervision of the chemical work and/or performing work involved in conducting special analyses and investigations, and directing chemical work; performing other engineering work as assigned; making appropriate recommendations for optimum plant performance; and making reports relating to his or her work.

Supervision:
1. Under direct supervision of the performance supervising engineer and/or under the supervision, direction, or guidance of the performance engineer-senior.
2. Give direction to employees assigned to him or her and assist in supervision, as assigned.

Responsibility and Authority: The performance engineer is responsible for and has authority to accomplish the following duties, as assigned, within the limits of company practices, policies, and procedures:

1. Perform and/or be responsible for work in the operating thermal efficiency of the plant, as assigned.
2. Plan, coordinate, supervise, and conduct test assignments, as assigned.
3. Test and inspect equipment, including boilers, turbines, heat exchangers, air compressors, generators, pumps, fans, air heaters, and other equipment.
4. Analyze and diagnose faults in the operation of instruments and control equipment and determine methods of correction.
5. Plan work and direct and supervise employees assigned to him or her.
6. Perform and/or be responsible for special analyses and investigations as plant conditions demand.
7. Assist in planning, scheduling, supervising personnel, and coordinating the work of the employees in the chemical work.
8. Assist in planning, scheduling, supervising, coordinating, reporting on and being responsible for operation of the chemical work and the necessary testing program of the components that go into the operation and maintenance of the plant, including coal, water, ash, lubricants, chemicals, and air.
9. Assist in administering policies, practices, and regulations of the plant as they relate to the chemical work.
10. Assist in supervising the use of the recording instruments and apparatus assigned to the chemical work.
11. Perform the other engineering work as assigned.
12. Prepare reports, analyses, and studies of his or her work in order to keep all interested people informed of the work he or she performs.
13. Make recommendations on changes for improvement in plant operation, equipment, material, etc., that result from analysis and study of the work he or she performs.

14. Train and instruct employees in safety regulations and performance of their work as assigned.
15. Maintain relationships with other personnel of the organization as are necessary to carry out functions.
16. Assume additional duties and responsibilities, as assigned, in the absence of the performance supervising engineer or performance engineer-senior.
17. Assume other duties and responsibilities, as assigned.

Qualifications:
1. Must be a college graduate in mechanical engineering, electrical engineering, chemical engineering, or chemistry; or have a professional engineer's license.
2. If progress from a performance engineer-B, must have several years of satisfactory work performance in that job and must have secured a professional engineer's license or a state professional engineer-in-training certificate.

Principal Lines of Promotion:
From: Direct employment of performance engineer-B.
To: performance engineer-senior.

An example of a person who would qualify for this position was 21 years old at the time of joining the company and had completed a college preparatory program in high school and an engineering degree in college. The person had held summer jobs in engineering-related situations and indicated on the application for employment an interest in electronics and a special aptitude involving analog computers.

Senior Clerk

The second example of jobs people hold in the electric power industry describes the position of senior clerk.

The position summary for this job is: "Performs, under *general* supervision, clerical duties which are more complex in nature, requiring independent analysis, exercise of moderately extensive judgment, and detailed knowledge of company and/or department work procedures. Performs complex calculations and has frequent contacts with other departments and some external contacts with customers, suppliers, and the general public, as required. Normally is assigned the *more* difficult clerical assignments and projects."

Position duties and responsibilities are in accordance with the company's procedures, practices, and safety regulations.

1. Carry out the following: review and process the more complex documents which require considerable procedural knowledge; assure completeness, accuracy, and compliance with company requirements.
2. Maintain records and files of these documents and reports in accordance with standard operating procedures.
3. Compile special reports, as required, from available records and request information from other areas when necessary.
4. Review reports, data, and information supplied by other clerical personnel, checking accuracy and adherence to standard procedures and policies and verifying or reviewing, in detail, the work of lower classified clerical personnel.
5. Answer customer or public inquiries, complete applications for service, process customer complaints, referring them to the proper personnel as required.
6. Answer and follow up routine complaints by telephone and/or letter, when assigned.
7. Initiate and/or answer routine correspondence, as required.
8. Act, when required, as senior ranking employee within organizational unit in absence of immediate exempt supervisor.
9. Carry out field checking duties when necessary.
10. Take and transcribe shorthand (and/ or machine) dictation as required.
11. Perform other duties as assigned.

Position Requirements (minimum necessary to perform job adequately): Education, training, or skills:

1. Perform required arithmetic calculations.
2. Read, write, and understand business English and grammar.
3. Perform such operations as post and check.

4. Adapt to office routines, and where required, have a general knowledge of shorthand.
5. Operate typewriter and one or more of the following equipment: adding machine, calculator, personal computer, computer terminal, word processor.
6. Have a valid driver's license as required.

Experience: The experience necessary for this position should normally include the progression through the classification of junior clerk, clerk, and intermediate clerk. This experience should be in line with or equivalent to the duties of the above job classifications. It is *generally* expected that the above progression to this position would be a minimum of five years. Where applicable, the incumbent must perform acceptably on the appropriate AEP System Placement Exercises, and must have demonstrated the ability to treat company business as confidential.

Line Mechanic A

In general, the duties of a line mechanic A are: "Perform all types of work involved in the installation, maintenance, rearrangement, operation, removal and inspection of energized or de-energized transmission and distribution facilities." Examples of specific duties follow:

1. Install and maintain circuits and all types of line equipment on congested construction energized or de-energized.
2. Select proper standard drawings and interpret specifications and proceed with all new construction work and live line work without supervision.
3. Supervise crew when assigned and display leadership and judgment when in charge.
4. Use live line tools and direct the operation of other employees when assigned.
5. Identify and connect additive and subtractive transformers in a three-phase bank.

6. Connect regular transformers to boost or buck voltage on a primary line.
7. Install and maintain series street light and multiple systems.
8. Direct the unloading of railroad cars and trucks.
9. Maintain tools, equipment and work areas in a clean and orderly condition.
10. Maintain required records.
11. Assist in the training of other employees.

Qualifications: Must have at least two years experience as a line mechanic B, or equivalent.

1. Must have sufficient knowledge and skills to perform the duties listed above.
2. Must have a good working knowledge of various circuit connections, voltages of lines, phasing and transformer connections, switching devices, ability to interpret field prints and switching diagrams, and to understand the fundamentals of electricity.
3. A good working knowledge of the tools and equipment used.
4. A good working knowledge of the company's transmission and distribution standards.
5. Ability to deal with employees, customers, and the general public in a courteous and proper manner.
6. Must possess a valid driver's license.
7. Qualification through demonstration, examination, and/or performance appraisal as determined by the company.

Principal Line of Promotion: From: line mechanic B to higher classification.

TRENDS IN EMPLOYMENT

There is much debate over increasing the demand for energy from all sources in the years to come. Some people emphasize how the increasing world population will require improving the efficiency of energy utilization rather than attempting to find new sources. Doing more with the same amount of electricity is seen as the best hope for avoiding socioeconomic problems, including the problem of maintaining the quality of the environment.

Speaking to this point of view, Con Edison, in a company-issued booklet called *Enlightened Energy,* says, "any kilowatt produced is still 'dirtier' than a kilowatt not produced. In other words, meeting demand for electricity by producing more of it will always have a greater environmental consequence than will avoiding the need for producing any additional supply." Consequently, there will be career opportunities for people who can increase the efficiency of electrical appliances, improve the effectiveness of insulation in heated or cooled buildings, reduce energy loss in transmission between power plants and final utilization, and develop programs that will balance out the demand for electricity.

It is to the industry's advantage to be able to meet peak demands without investing in more than the minimum amount of generating capacity. To gain this advantage, one program gives customers credit for agreeing to "time-of-day" reduced consumption during periods of peak demand. Another program encourages customers to use electricity during periods of low demand to chill or freeze water and store it to air-condition buildings during periods of peak demand.

To be weighed against the possible trend toward decreasing energy consumption is the fact that there is a shift away from other energy sources toward electricity. Quoting the United States Department of Energy's 1983 report on "The Future of Electric Power in America," there will be a "substantial increase in the electric intensity of the economy and an increase in the share of electricity in the delivered energy market. Even though higher prices may somewhat abate the demand for energy, the economy is very likely to turn increasingly to electricity for its energy requirements."

It follows from this prediction that employment opportunities will increase in the electric power industry. Furthermore, if the industry is to expand sufficiently to insure that peak demands can be met, problems involving environmental pollution caused by burning "dirty" fuels, routing transmission lines through urban areas, and scheduling new constructions must be solved. Many of these problems fall into political and economic categories and will appeal to people who like to wrestle with political, sociological, or economic problems.

Research Opportunities

Ongoing research within the electrical power industry also provides employment opportunities. The goals of research can be as clear-cut as looking for ways to keep trees from interfering with power lines. A sizeable item in the annual maintenance budget for power distribution systems is the cost of trimming trees to keep them clear of overhead power lines. The cycle for retrimming trees, usually every one or two years, depends on their growth rate. This leads to the question of how to reduce growth rates without damaging the tree.

Work already done suggests that a chemical can be found that will reduce the growth rate of the sucker sprouts which appear on newly pruned trees. This chemical can be injected directly into the trunk of the tree or applied as paint in a band around the tree trunk. Thus trees that had to be pruned each year can now be put on a three-year trimming cycle.

Another research topic involves improving ways to transmit electrical power through underground cables. Efficient transmission of power to the centers of metropolitan areas demands underground transmission. Not only do homeowners complain about the appearance of overhead transmission lines, but overhead lines require rights-of-way hundreds of feet wide. Even if the land were available, its cost would add immensely to the cost of the electrical power transmitted.

Underground transmission lines are made by wrapping many layers of paper tape impregnated with mineral oil around the metal cable. Layer after layer of oiled paper produces a uniform insulation as much as an inch thick. This insulated cable is snaked through rigid metal pipes which are then filled with oil under pressure. Even the tiniest bubble or wrinkle in the layers of insulating paper can lead to loss of power or ultimate failure of the underground cable.

To repair an underground cable isn't easy. Once the trouble spot has been located and the pipe uncovered, a section on either side of the trouble spot is frozen to immobilize the oil. Then the oil is drained from the portion to be replaced. To replace the insulating paper and to weld a new length of pipe in place may take a three-shift crew several weeks of round-the-clock work.

Research projects are looking for new plastic tapes to replace the paper wrapping. Bolder research is aimed at using a plastic that can be extruded to form the insulating layer rather than wrapping the metal cable with paper or plastic tape. As yet, however, the thick layer of insulating plastic cannot be extruded to form a sufficiently uniform thick layer around the metal cable.

Researchers with even longer time budgets are looking for totally new materials to use in making underground transmission lines. One intriguing idea is to replace the usual copper or aluminum metal with sodium. This metal is light, cheap to produce, and flexible. It is not as good a conductor as copper or aluminum, so larger cables would be called for. More serious difficulties are that sodium melts at a temperature below that of boiling water and reacts violently when exposed to moisture. Despite these disadvantages, polyethylene-covered sodium cables are being tried experimentally.

The electric power industry is the direct link between nearly all people and our nation's major energy resources. True, electricity as such is not an energy resource per se. It becomes available to us only by transforming some other energy resource into electricity. If these other energy resources come to be in short supply or too expensive, the resulting problems will be brought home to us most painfully by way of the electric power industry. On this basis, there is little doubt that jobs and careers within this industry will become more and more important as the years go by.

POWER PLANT JOBS AT A GLANCE

Engineers and Scientists
(Generally requires a four-year bachelor's degree and often requires postgraduate study)

electrical engineers	nuclear engineers
power engineers	nuclear physicists
mechanical engineers	environmental scientists
design engineers	civil engineers
environmental engineers	

Support Workers
(Generally requires at least a two-year associate degree for technicians; technologists require a four-year bachelor's degree)

engineering technologists	laboratory technicians
engineering technicians	environmental monitors
	computer operators and programmers

Blue Collar Workers
(Most jobs require apprenticeship. Some may require special nuclear certification)

supervisors	ironworkers
pipe fitters	boilermakers
electricians	painters
carpenters	sheet metal workers
welders	bricklayers

Semiskilled Workers
(May require a high school education)

truck drivers	laborers
equipment operators	

CHAPTER 9

ENERGY FROM THE EARTH'S NONDEPLETABLE SOURCES

Winds, rivers and waterfalls, ocean currents and tides, and the heat in volcanoes, hot springs, and geysers are energy resources that offer uniquely inviting career opportunities. The winds will always blow and rivers will always flow. The tides are eternal, and ocean currents will always move enormous quantities of water from warm to colder areas. It is difficult to believe that volcanoes and hot springs will ever totally dissipate the heat in the rocks deep in the earth's crust.

Accurate predictions of the number of job openings that these energy sources will provide must take into account changes in the availability and cost of more readily developed energy resources. It is more a matter of economics than the quantity of energy that is involved. It is almost impossible for a business enterprise to justify investing the money needed to develop and distribute the energy in these nondepletable sources.

On the positive side, it is reassuring to know that before natural gas, oil, and coal were widely available, waterwheels and windmills gave muscle power much support. Furthermore, these energy sources lend themselves to helping individuals become "energy independent."

WIND POWER

The 1989 budget of the U.S. Department of Energy included nearly $9 million to support research that might decrease the cost and improve the efficiency of windmills, which are more appropriately called wind turbines. Research projects looked for improvements in the design of blades and in the machinery that transforms wind energy into electrical power. Basic research looked into how wind turbine blades interact with the wind and how to locate windmills where the greatest amount of energy can be harvested.

How steadily the winds blow and at what speed determines how much energy can be obtained. Each location also may have topographical features that either concentrate or disperse the wind. Fortunately, records kept by governmental agencies often include the average daily windspeed for many locations, particularly at airports. Areas particularly well supplied with potential wind energy become "wind farms." Examples are the Altamount Pass near San Francisco, the Tehachapi Mountains north of Los Angeles, and several sites in Hawaii.

Large wind power systems are connected to the public utility's power lines. When the wind system generates excess electricity, the excess can be sold to the utility. In turn, when the wind isn't blowing hard enough, customers can fall back on the utility's electricity. Improving the sociopolitical circumstances that will enable wind power to gain increased acceptance, as well as solving economic problems, will make careers in this area rewarding.

Where It Comes From

Wind energy comes from interactions between the sun's energy falling on the earth, and earth's gravitational forces. Because the earth rotates and revolves around the sun, different areas of the earth receive unequal amounts of sunlight during different times of the day or seasons of the year. This causes the air to be heated unequally. When air masses are heated or cooled at different rates, they expand or contract unequally. With unequal changes in volume, the earth's air masses develop different densities. Gravitational forces cause the more dense masses to settle and the less dense masses to rise. Winds are the earth's air masses in motion.

To illustrate this complex set of interacting forces, think of a northern lake still covered with winter ice and snow. When spring comes, the sun's energy melts the snow and warms the countryside more rapidly than it warms the surface of the lake. In turn, the air above the land areas becomes warmer than the air above the lake. The warm air above the land will rise and the air above the lake will move to take its place, with the consequent exchange of energy.

A somewhat different situation may occur later in the spring or during the summer. During daylight the land areas may be warmer than the lake's surface. But when the sun sets, the air above the land areas cools more rapidly than does the air above the lake. In turn, the air above the land areas will move to replace the air that rises above the lake.

Tornadoes, tropical storms, and other types of destructive windstorms emphasize that the earth's winds pack an enormous amount of energy. In more pleasant settings, the gentle pull of the wind on a high-flying kite or the quiet urging of the wind-filled sails of a sailboat provide examples of wind energy being put to use.

History of Wind Power

Between the 1400s and 1800s, much skill was invested in designing machines that would capture the wind's energy. In time, windmills became a part of the landscape in Holland, Belgium, and the newly settled, windswept plains of the western United States.

The European type of windmill usually had a small number of very large canvas-covered vanes or sails. In America as many as twenty or more small metal vanes were used to catch the wind. In both types, cleverly engineered devices kept the vanes head-on into the wind. Other devices adjusted the pitch of the sails or vanes to fit the speed of the wind and the amount of work to be done by the spinning windmill. Gear systems protected the windmill during excessive winds and transferred the energy of the whirling vanes to where it was needed to pump water, grind grain, or do other useful work.

The windmill industry provided more job opportunities between 1850 and 1930 than during recent years. The industry was pretty much put out of business by the extension of electricity to rural areas throughout the United States. It is difficult to guess how rapidly the rising cost of other sources of energy might revive the demand for wind machines.

Several large industrial organizations are taking on research and development programs, and the U.S. Department of Energy has sponsored the development of several types of wind systems. Small machines capable of producing enough wind energy to drive generators that yield five to ten kilowatts of electrical power are already at work. These machines would supply the power needs of homes, farms, or ranches. The department is also looking into the possibilities of tying many of these small units together so that their output can be pooled to provide large blocks of power for towns and cities.

Wind turbines with blades up to three hundred feet in diameter are also being designed. These machines are expected to be able to feed electricity into existing power systems and either supplement or replace fossil fuel-powered electricity generating systems.

Employment Trends and Opportunities

Again, it is difficult to predict trends in the number of employment opportunities that will occur in the wind industry. We know that a family can build a windmill that will convert enough wind energy to electrical energy to meet many of the family's energy needs. Storage batteries can be used to store energy to tide the family over during windless periods. We are less sure that wind power could replace central power stations which generate electricity by burning coal or oil.

It is not so much that there isn't enough energy in the earth's winds. It is more a problem of harvesting the energy when the winds blow, storing the energy when it is available in excess, and making it available as needed whether or not the wind is blowing. One proposed solution to this problem is rather simple: Use the electricity that is available when the winds blow to separate water into its components, hydrogen and oxygen. Stockpile the hydrogen and oxygen. On calm days, fall back on generating electricity by feeding these gases into fuel cells that will convert the energy released when the gases react to form water. Use this energy to drive steam turbines or, better yet, allow the fuel cells to produce electricity directly.

To improve the efficiency of wind systems and the methods for storing energy points toward the kinds of challenges that await people who look forward to jobs or careers in the energy industry.

Studying Wind Power

The Westinghouse Corporation received a government contract to spell out the problems that must be solved before the energy available in the winds that sweep over oceans and lakes can be used to help head off an impending energy shortage. The assignment given to Westinghouse shows clearly the career opportunities that will develop if this energy resource is fully exploited.

One part of the assignment was to identify sites offshore where wind power is especially available. It is well known that the winds blow more uniformly over bodies of water than over land. Wind velocity over the oceans and large lakes is also higher than over similarly located land areas. And the speed of the wind has much to do with the amount of energy that can be harvested from wind power.

The energy available to a windmill or turbine increases with the cube of the wind velocity. A thirty mile per hour wind packs twenty-seven times more energy punch than does a ten mile per hour wind. From another point of view, a turbine would receive as much energy from a thirty mph wind in one hour as it would from a ten mph wind in twenty-seven hours.

The Westinghouse study also examined the cost of transmitting energy, primarily in the form of electrical power, from an offshore wind power facility to where the energy is needed. Other parts of the assignment involved designing a structure to support wind-powered turbines, predicting the effect of windpower facilities on the environment, examining various ways to transport the energy that is collected by windmills and turbines to where it is needed, and predicting the cost of the energy that might be collected from the winds that blow over the oceans and large lakes.

Offshore wind energy collecting systems involve pretty much the same engineering problems that the oil industry faced when it began offshore drilling operations. Towers for the wind generators must be built either on foundations that rise from the ocean bottom or on floating platforms. Environmental problems are also similar. Critics are sure to question how wind power systems will affect recreational facilities, commercial fishing, shipping lanes, and so forth.

The fact remains that there are areas off the shore of the United States where the winds blow regularly and at relatively high velocity. To harvest the energy that these winds produce challenges the ingenuity of people who are interested in making sure we have adequate supplies of energy.

WATERPOWER

Waterpower at one time provided a much appreciated solution to energy problems. To provide waterpower, whenever possible, towns and villages were settled along streams, and "down by the old millstream" was more than a nostalgic expression.

Modern hydropower facilities tap the energy of falling water at dam sites to drive turbines that generate electricity. Such facilities can be as large as those at Niagara Falls and Grand Coulee Dam or as small as unnamed dams built at mountain retreats to provide electricity for hideaway vacation spots.

The energy of falling water can be traced to the circulation of water through the atmosphere and hydrosphere. The sun evaporates water from the oceans, lakes, streams, and the moist earth. The vapor-filled air cools as it rises. If the air becomes cool enough, the water vapor holding capacity of the air is exceeded and condensation occurs. Rain or some other form of precipitation falls in response to gravitational forces. These same forces keep the water moving until it approaches as closely as possible the center of the earth.

Thus, the energy of waterpower comes from interactions between solar energy and the earth's gravitational forces. This means that waterpower is an energy resource that will be available as long as the sun shines and the rains fall.

To convert the kinetic energy of falling water to electric power requires a dam or some other structure that makes it possible for a head or store of water to be maintained above a downstream channel. As much of the water as is needed to turn the turbines which drive the generators is directed by way of pipes so that it must spin the turbines as it falls. The excess flow of water is allowed to go over a spillway as it continues its way downstream.

In addition to the job opportunities that exist in the operation and maintenance of hydropower facilities already in operation, additional opportunities may develop. Hydroelectric plants now account for about 14 percent of the total electricity being generated in the United States. Less than one-fourth of the nation's available waterpower has been developed. Worldwide, if all waterpower resources were to be developed fully, they could generate five or six times as much electricity as was being generated worldwide in 1970.

A somewhat conflicting point of view says that most large hydroelectric sites in the United States have already been developed. A survey by the Army Corps of Engineers, however, has identified fifty thousand existing small dams that might be adapted for hydroelectric generation. Obviously, if these possibilities were to be exploited, many new job and career opportunities would be created.

If dams and their accompanying lakes are built early in the life of an area, few environmental problems are involved. Not so when a dam is to be built in an area where the new lake would flood large sections of an established community. An untamed river crashing down the cataracts and rocky gorges of scenic mountain country can be a valuable recreational resource. It can also be the site of a hydroelectric facility.

With proper planning, the recreational and hydroelectric resources, as well as other valuable land-use purposes, can be developed without destroying too much of the original resource. Important career challenges await people who can deal with the situations in which our energy needs interact with other needs and interests of society.

Improving the efficiency of waterwheels and turbines provides research career opportunities. The overshot and undershot wheels of colonial America were scarcely 50 percent efficient. Turbines can be as much as 95 percent efficient. Reaction turbines are highly efficient systems which depend on pressure rather than velocity to transform the energy of flowing water into mechanical energy. These turbines are placed in streams of moving water where pressure causes the blades to rotate.

Interest in hydropower, as in other nondepletable energy sources, is increasing as people become increasingly concerned about maintaining environmental quality and becoming more and more dependent on imported oil.

ENERGY FROM OCEANS

Exceedingly challenging and potentially rewarding jobs and careers await people who are willing to apply their minds and hands to harvesting energy from the oceans. Enormous amounts of energy await being harvested from ocean currents, tides, waves, and differences in the concentration of dissolved salts.

The oceans' tidal systems could produce as much as 12,000 trillion kilowatt-hours of electrical power. This is enough power to drive the whole world's industrial machinery. Of course, this estimate assumes that tomorrow's engineers will learn how to harness the tides' energy with total efficiency. Although the goal of total efficiency is unrealistic, there is enough energy in the ebb and flow of the earth's oceans to catch the imagination of creative people who are looking for new sources of available energy.

Like all phenomena that involve gravitational forces, the coming in and going out of oceanic tides are hauntingly mysterious. It takes a good imagination to see the earth's oceans as enormous flattened bubbles of water sticking to the earth's surface. This means, of course, that when viewed from a vantage point in space outside the earth, some of these bubbles appear to be tilted or even "upside-down."

Gravitational forces cause the water in these oceanic bubbles to stay as close to the earth's center as possible. But the earth's gravitational forces are not the only forces acting on the oceans. The gravitational forces of the moon, the sun, and all other objects in the universe threaten to "spill" the water from the earth's oceans. Only the moon and sun have mass enough to compete noticeably with the earth's own gravitational attraction for its oceans. Although the moon is many times less massive than the sun, it is so much closer to the earth that its gravitational forces are primarily responsible for the "sloshing to and fro" of the oceans.

Each day, because of the earth's rotation, the oceans revolve eight thousand miles closer to, or farther from the moon. Variations in the moon's gravitational attraction for the water in the oceans in relation to the earth's gravitational forces causes the oceans to spill slightly toward or away from the moon.

Coastal irregularities sometimes allow the incoming tide to impound water that can escape only through a narrow channel when the tide goes out. Dams or barriers can be built across these narrow channels. The barriers are opened during incoming tides and closed when a pool of water has been trapped. The trapped water can then be made to drive turbines and generate electricity before it can return to the level of the ocean during low tide.

One such situation is the Bay of Fundy, bordering on Maine, Nova Scotia, and New Brunswick. Irregularities in the Atlantic coastline create more than five hundred square miles of tidal pools. Differences between high and low tides run as high as thirty feet. Engineers estimate that there is enough tidal energy here to produce electricity sufficient to meet the needs of 600,000 people. There are equally promising tidal power sites on the coast of France.

Another set of closely related career opportunities awaits people with truly superior imaginations. An American physicist, Albert A. Michelson, proved that the earth's surface—with all of its cities, hills and mountains, highways, and forests—rises and falls periodically. It is not only the oceans that ebb and flow under tidal forces.

There are also earth tides which involve tremendous amounts of energy—amounts of energy that may invite more and more attention when other sources of energy appear to be dwindling.

Energy from Ocean Waves

To feel the overpowering force of oncoming waves when swimming in the relatively mild surface of a seaside summer resort is sufficient evidence that ocean waves carry an enormous amount of energy. More than 1 percent of the energy that comes from the earth from the sun is transferred to the energy of the winds. This is the energy that causes the relentless motion of the surface of the sea, and this energy source has long captured the imaginations of men and women who believe that there should be ways to put this energy to useful work.

Waves have been known to come crashing toward the shore with forces as great as six thousand pounds per square foot. On the Oregon coast a twenty-ton rock was hurled by the Pacific's waves toward the shore with such force that it damaged the roof of a lighthouse. On the coast of Scotland, a 1,350-ton block of concrete was broken loose from a breakwater by wave action, and a few years later its 2,600-ton replacement was literally washed away.

Many ideas for capturing this enormous energy have been tried, but none has proved to be successful on a large scale. Wave energy is used to blow whistles and ring bells on ocean buoys. But to design a wave-driven power system remains a challenge to the engineers of tomorrow who are willing to match their talents with this undeveloped energy resource.

People with special mechanical aptitude have designed several kinds of wave motors. A system of levers and wheels attached to large floats transformed the up- and down-motion of the floats into forces that were used to turn drive shafts. Similar mechanical devices used the sloping surfaces of waves to cause floats to slide down the face of a wave and thereby operate a wave motor. The crashing force of a wave rolling across the shore has also been converted into the energy input for small wave motors.

In slightly more sophisticated approaches to harnessing ocean waves, a concrete slope forces incoming waves to deliver water to a storage basin well above the usual ocean level. Before the water can run back to the ocean, it must flow through a turbine and spin a generator which converts the energy initially in the waves into electrical energy.

JOHN ISAACS

John D. Isaacs and his colleagues at the Scripps Institution of Oceanography in California have devised another way to force the waves to lift water so that it can be made to do useful work when it returns to the level of the sea. A large, long pipe hangs suspended from a float. When caught in the trough of a wave, the pipe with its float sinks and the pipe fills with water. But when an incoming wave lifts the float and the partially water-filled pipe, a check-valve in the bottom of the pipe keeps the water from flowing from the pipe. The approach of another trough lets the pipe sink and a new quantity of water enters the pipe.

Repeated rising and falling with the ups and downs of more and more waves fills the pipe until the water overflows and is collected in a storage tank well above the ocean surface. Before the water can leave the storage tank and return to sea level, it is made to generate electricity or do other useful work.

YOSHIO MASUDA

Rather than using a floating pipe to capture and compress the air, Yoshio M. Masuda designed an eight-sided ring-shaped buoy 120 meters in diameter and 4 meters thick. When the structure floats on water, air is trapped underneath. Oncoming waves lift one side of the structure and water rushes in to force the air toward the opposite side. The air cannot escape, however, without passing through a turbine that drives an electrical generator. As the moving wave reaches the opposite side of the floating octagonal structure, this side is lifted while the original side drops down into the trough of the wave. The alternating lifting and dropping of opposite sides of the structure, together with cleverly designed valves, feeds a stream of compressed air into the turbine. Masuda expects this structure to yield six thousand watts.

MICHAEL McCORMACK

This type of wave machine has sparked the inventiveness of Michael E. McCormack and his team of research engineers at the U.S. Naval Academy. Their research says that such a wave machine should be especially attractive to the people who design and build offshore drilling platforms as well as other offshore installations. By using a large pipe in the floating center and arranging to have the length of pipe submerged adjustable to the height of the waves, McCormack's team believes that this type of wave machine could harvest more than enough energy to provide the power requirements of offshore installations, and might even serve customers onshore.

Several years ago, Bouchaux-Praceique built a shore-based wave power facility that generated all of the light and power needs for his seaside home at Royan, France. A tunnel ten feet below the lowest tide level was connected to a vertical shaft outside the seawall. The top of the shaft was sealed airtight but it was fitted with a pipe that allowed air to flow to a turbine. Waves crashing into the tunnel caused the air in the shaft to be compressed or decompressed with the incoming and outgoing waves. Very clever engineering forced the turbine to spin when either high or low pressures developed in the shaft.

The advantages of this installation were that no metal parts were exposed to the corrosive seawater, there was no chance for floating seaweeds to clog piping, and the destructive force of storm-size waves was absorbed by the natural rock seawall.

STEPHEN SALTER

Stephen H. Salter at the University of Edinburgh tackles the head-on energy of oncoming waves in the design of his wave energy converter. He designed a structure with a partially submerged vane that faces directly into the oncoming waves. The shape of the vane is suggested by the informal name, "nodding duck." The leading edge is shaped so that the "duck" rotates against the force of the oncoming wave but the trailing edge allows the water to "roll off the duck's back."

In a full-scale installation, a string of the ducks would be connected along a tubular backbone that would allow each duck to rotate individually. The energy absorbed by all of the ducks would be picked up by levers and gears within the backbone. In turn, the collected energy would be used to drive motors or turbines for generating electricity.

Salter foresees a commercial version of his idea to consist of several parallel rows each the size of a supertanker. These strings of ducks would be loosely moored at sites where wave motion is consistently available and as close as possible to cities where there would be a market for power.

CAREER OPPORTUNITIES IN WAVE ENERGY

These descriptions of inventions intended to harness wave energy bring out something that is especially meaningful to people who are thinking about building careers in the energy field. This field continues to encourage and reward downright inventiveness and creativity. The person who likes to tinker around in a workshop or design studio can find an important career niche in the nation's concern for its energy supply. Obviously, large amounts of money are needed to build and try out large-scale models, but this doesn't mean that large amounts of money are needed to think through large-scale ideas.

Running parallel to the development of installations to harvest wave energy is the problem of environmental impact. This problem also creates career opportunities. Some people believe that floating structures designed to harvest wave energy can be designed in ways that will attract fish or improve the environment for marine life. Other people believe they should devote their time and talents to protecting marine environments by proving that installations intended to harvest wave energy should be prohibited. Valuable careers can be built when people set out to inform the public accurately and effectively about the need for wave energy-harvesting installations and the pros and cons of their impact on the environment.

Energy from Ocean Currents

In 1978, the Congressional Research Service of the Library of Congress prepared a 433-page report titled "Advanced Energy Technologies and Energy Conservation Research, Development and Demonstration." The report was prepared primarily for the use of the Congressional Committee on Science and Technology.

To introduce the topic of energy from ocean currents, the authors quote Matthew F. Maury's wonderful description of the Gulf Stream, one of the most powerful currents in the world's oceans:

> There is a river in the ocean. In the severest droughts it never fails, and in the mightiest floods it never overflows; its banks and bottom are of cold water, while its current is of warm. The Gulf of Mexico is its fountain, and its mouth is in the Arctic Seas. It is the Gulf Stream. There is in the world no other such majestic flow of waters. Its current is more rapid than the Mississippi or the Amazon, and its volume more than a thousand times greater . . . The Gulf Stream is one of the most marvelous things in the ocean . . . he who contemplates the sea must look upon it as a part of the exquisite machinery by which the harmonies of Nature are preserved.

A portion of the Gulf Stream flows between southern Florida and the Bimini Islands. Called the Florida Current, some thirty million cubic meters of water flow through this fifty mile wide "river" each second. The average velocity is slightly less than a meter per second, but under certain conditions the velocity builds up to more than 2.5 meters per second.

To sense what this means, a ship moving against the Florida Current must call upon its engine to create enough power to equal that needed to maintain a forward speed of about five knots simply to stay where it is. Looked at from another point of view, if the ship were anchored and its drive mechanism were to be driven in reverse by the flow of the Florida Current, the Current would generate as much power as is generated by the ship's engines when under way at a speed of five knots.

Obviously, the anchored ship would intercept only a small fraction of the massive amount of flowing water. No wonder the total energy represented by the Florida Current and other similar currents in the oceans in other parts of the world challenges the minds of many people—especially people who are seeking career spots in the energy industry.

As yet nobody has built an installation that will harvest energy from ocean currents. But ideas are being generated. Most schemes begin with pretty much the approach of an anchored ship or floating structure with a turbine driven by the current. Turbines are usually built to take energy from a closely confined flow of water. Turbines designed to take energy from ocean currents, however, are more likely to be immersed in the flowing water and the blades made to spin only by the water that the blades intercept.

In effect, an underwater turbine to take energy from an ocean current is like a windmill designed to take energy from the passing wind. The turbine can be built within a housing somewhat like the housing that surrounds an airplane jet engine, or the blades of the turbine can be left exposed. Just as windmills must be built with a mechanism to keep the blades flat against the wind, underwater turbines must be attached to structures that can swing to adjust to changes in the direction of the ocean current.

Several people are working on ingenious devices that harvest the energy of ocean currents more efficiently. Gerald F. Steelman has invented a "Water Low Velocity Energy Converter." His idea involves a simple pulley wheel joined to a vertical shaft which operates a generator and is held in position by an anchored ship. Around the pulley is a loop-drive belt along which are spaced parachutes or "sail canopies" designed to open and catch the flow of water when moving against the current and to collapse when moving with the current.

Steelman foresees a system consisting of parachutes one hundred meters in diameter spaced along a drive belt ten or so miles long. This system would intercept a working volume of water weighing 120 million tons. The energy collected would be used to separate water into its elements, hydrogen and oxygen, and these fuels could then be transported to shore and sent to market. In some situations, the installation could be built close enough to shore to allow the captured energy to be converted to electricity and sent to shore through underwater cables.

Other inventors propose building a set of attached turbines in honeycomb fashion, mounting them on an underwater platform, and collecting the energy generated by the turbines from the passing water. The entire installation could be far enough below the surface to avoid being a hazard to shipping and to be protected from hurricanes and other storms.

Again, these efforts to harvest energy from ocean currents are not far enough along to create clear-cut job and career opportunities for large numbers of people. But the courage that is reflected in such exploratory efforts is to be admired. We will always be able to solve our energy supply problems so long as people with this kind of courage and vision seek careers in the energy industry.

ENERGY FROM OSMOSIS AND PHOTOSYNTHESIS

School biology students learn that water is forced through a permeable membrane that separates two solutions of different concentration. It is difficult, however, to see how this principle could be used to harvest energy from rivers and oceans.

We also learn that photosynthetic processes in the leaves of green plants store solar energy in the sugars and starches and other carbohydrates that are put together by these photosynthetic processes. We know that green plants produce fuel in the form of wood or wastes from farm crops. Could it be that the oceans are a source of plant materials in large enough quantities to make a significant contribution to the nation's energy resources?

Efforts to use osmosis and photosynthesis to solve our energy supply problems vividly illustrate one feature of energy career opportunities. Every situation in which energy is stored can be the target of research. People who are looking forward to energy-related careers become particularly sensitive to and aware of any process in nature that tends to store energy. Each such situation is then examined from the point of view of the specific scientific principles which are involved. Having identified these principles, ways are then looked for that will make the stored energy available as a contribution to the nation's energy resources.

People whose only contact with osmosis is seeing wilted celery stalks regain their crispness after standing in cold water may be surprised to learn that osmosis is being used to make new sources of energy available. We know that osmosis can force water to rise in wilted celery stalks. But it takes a good imagination to think of osmosis being used to harvest energy from rivers and streams.

Osmosis Research

Gerald Wick and John Isaacs of the Scripps Institution of Oceanography calculate that osmosis can be used to take as much energy from, say, the Columbia River as is produced by the hydroelectric installations on the Columbia. Behind their calculations is the fact that river water contains lower concentrations of dissolved salts that does ocean water. The fluids in the cells of wilted celery stalks also contain higher concentrations of dissolved substances than are present in cold water. The cell walls of the celery create semipermeable membranes, and osmotic pressure forces water to rise in the stalk.

Wick and Isaacs argue that if a semipermeable membrane could be stretched across the mouth of a river where it meets the ocean, osmotic pressure on opposite sides of the membrane would develop enormous force. This force could be used to lift water into a storage tank. In fact, Richard Norman of the University of Connecticut says that osmotic forces at the mouth of a river create the equivalent of a storage dam 225 meters high. The ability to see such a massive structure in the invisible forces of osmosis is characteristic of the unique capabilities of the people who tackle our energy supply problems.

In general, efforts to use osmosis to harvest energy fall somewhere between accurate description of the scientific principles involved and the design of small-scale laboratory models or pilot projects. Several types of systems are being studied. One system uses an artificial lake in which fresh water is stored. This lake is built several hundred meters below the level of the river that feeds the lake.

A hydroelectric power station takes energy from the water as it falls from the river into the lake. Still more energy is harvested when the water must pass through a permeable membrane before it mixes with ocean water. This is done by using osmosis to put the water under pressure; the pressurized water then spins turbines which drive generators.

Sidney Loeb of Ben-Gurion University of the Negev in Israel is looking into a system where water flows by osmosis through a series of semipermeable membranes into a pressure chamber. Pressure developed in the chamber will then force water through an exit with sufficient force to spin turbines.

Octave Levenspiel and Noel de Nevers at the University of Utah designed a system involving a large vertical tube immersed in seawater with a semipermeable membrane on its lower end. The tube remains empty until the pressure of the water above exceeds the osmotic pressure. At this point, water flows through the membrane in a direction opposite to that of normal osmotic flow. Because fresh water is less dense than seawater, the water would rise in the tube above the level maintained by osmotic pressure. When the fresh water is allowed to return to the level of the seawater, it is forced to spin a turbine.

John Weinstein and Frank Leitz at the National Institutes of Health and the Department of the Interior have studied a system consisting of a series of specially designed membranes and electrodes. These special membranes allow positive ions such as sodium to build up in one compartment and negative chloride ions to build up in another compartment. When the charged particles collect in separate compartments, a flow of electric charge or current is generated. In theory, sufficiently large quantities of charge could be produced to create marketable quantities of electricity.

The wide array of ideas that have been sparked by the energy that is still largely latent in the osmotic forces between fresh and seawater bear witness to the satisfactions people derive from matching their wits and hands against the challenge of solving our energy supply problems.

Photosynthesis Research

People who are especially attracted by photosynthetic processes as possible solutions to our energy supply problems point out that more than 70 percent of the earth's surface is covered by the oceans. They also call our attention to the lush kelp beds which form underwater "forests" near the shores of oceans in many parts of the world. By putting these facts together, some scientists foresee harvesting the end products of photosynthetic processes going on in oceanic environments and converting these products into fuels.

A project developed jointly by the United States government and the American Gas Association goes one step further. In this project a kelp farm provides a continuous supply of raw material. Processing plants included in the project would convert the harvested kelp into synthetic natural gas, alcohol, fish and cattle food, fertilizer, and other saleable products. For additional income, the fish that would be attracted to the kelp farm would be caught and sent to market.

The economics of such a kelp farm and processing project are yet to be worked out. Perhaps such a project can take advantage of one of the ways to harvest energy from ocean currents, waves, or sea solar energy. This would reduce processing costs. For sure, energy storage by photosynthesis is cheap and reasonably efficient. There is no doubt that people can gather their own kelp or other biomass fuel and use it to solve their own energy supply problems. It is another matter when kelp farms or other versions of the original "farm woodlot" are to be used to provide enough energy to meet the needs of people in large cities.

BIOFUELS AS A SOURCE OF ENERGY

Biofuels take us a step closer to realizing how dependent we are on solar energy. Biofuels come from things that have been alive, and life depends upon photosynthesis whereby the sun's energy is stored in plant tissues and organs throughout the biosphere. Coal, oil, and natural gas are fossil forms of biofuel and their origins were equally dependent upon collecting and storing solar energy.

Biofuel and muscle power provided nearly all of the energy that kept America going during its birth and adolescence. But when fossil fuels were discovered and developed, wood and other biofuel was pretty much abandoned as an energy resource. It is worth noting, however, that wood and other biofuel regained 3.7 percent of the energy market during the 1970s when adequate supplies of oil and gas were difficult to obtain.

Biofuel can be used to produce electricity, heat, and vehicular fuel. Its source is usually waste material left over from various endeavors. Typical examples are forest products such as sawdust, bark, scrap lumber, and paper; fruit pits, nut shells, corn cobs, sugar cane bagasse, and manure from agriculture and food processing industries; and domestic and industrial sewage and solid wastes. It also makes good sense to salvage the energy in the trash left by the timber industry and in the residues from farm crops.

If the cost of other energy resources makes it economical to do so, farmers can turn to producing high-yield, high-energy crops such as sugar cane, sugar beets, and sorghum or rapidly growing trees such as eucalyptus, willow, and sweet gum as cash crops grown specifically as biofuel.

Career opportunities are sure to increase for people who are interested in finding more efficient ways to convert the stored energy to more usable form. Burning biofuel to produce heat is the most commonly used conversion process. Gasification processes heat biofuel in the absence of oxygen. Also called destructive distillation, these processes convert plant material to gases, liquids, or solids which, in turn, can produce substitutes for oil and natural gas.

Microorganisms have long been used to break complex hydrocarbons and carbohydrates into simpler substances in which the stored energy is more readily available. Using yeasts to convert carbohydrates in grains and fruits into ethyl alcohol may well be humanity's earliest chemistry. Equally long-standing is the use of microorganisms to convert organic wastes into plant nutrients. More recently, sewage disposal systems rely on microbes acting in aerobic environments to not only get rid of waste but to also produce methane, which is the principle component of natural gas. In California, a dairy processing plant solves its energy needs by collecting methane that is produced by anaerobic digestion of dairy waste products. Other dairy operators have developed manure handling systems in which anaerobic processes produce sufficient fuel to generate enough electricity to not only meet their demands but yield a surplus to be sold to local utilities.

As will all energy sources, biofuel development can be both environmentally favorable and unfavorable. On the negative side, essential plant nutrients are not returned to the soil but are scattered about. At the same time, properly managed biofuel utilization can solve waste management problems in ways that are environmentally beneficial. Opportunities to maximize the positive and minimize the negative features of biofuel development and utilization makes careers in this area doubly attractive.

CHAPTER 10

ENERGY CAREERS: SOME COMMON THEMES

There are special satisfactions to be enjoyed in all energy-related careers. There are also special responsibilities. This is true without regard to wages or salaries or how many years of training are required. It is also true whether one's job reaches no farther than neighborhood utilities or extends into the ocean depths or outermost space.

The satisfactions come from enabling energy to play its essential role in everything people want and need to do. They also come from allowing energy to play its equally vital role in everything that happens in nature, and particularly in helping to maintain an environment that sets no limits on humanity's hopes and ambitions and scale of values.

The responsibilities can be as simple as what it takes to get one's job done, or as far-reaching as protecting and improving the quality of life that is available throughout the global community. Some energy-related jobs put people close to large quantities of readily available energy—energy that if handled properly can do good things for humanity, but if mishandled, can blow the surroundings to smithereens. There is no place for error or misjudgment when people work with flammable motor fuels, place explosive charges that are to dislodge tons of rock in coal or uranium mines, adjust the fuel rods that control the rate of nuclear reactors, install leak-proof gas lines, or throw switches that control high voltage electrical circuits.

SOCIOPOLITICAL IMPLICATIONS

The Persian Gulf War of 1990–91 emphasizes how the availability of adequate supplies of energy adds a whole new dimension to the responsibilities of the people who make a nation's sociopolitical decisions. During this war, 732 Kuwaiti oil wells were deliberately set afire by Iraqui troops. Estimates of the amount of oil gushing from these burning wells ranged from 1.5 million to more than 6 million barrels per day. Estimates of how long it would take to extinguish the fires reached several years into the future. However, all fires were extinguished before the end of 1991. The carbon dioxide, soot, and other noxious gases and solids created sun-blocking smoke plumes that rose twenty thousand feet and were caught up in the worldwide circulation of the atmosphere. As a result, this one event became a massive contribution to the environmental pollution problems that the global community had hoped to bring under control before the end of the 1900s.

Viewed from the perspective of making the most possible from nature's energy resources, the burning of Kuwait's oil wells is a catastrophic lesson from which humanity can learn the possible negative consequences of sociopolitical decisions that fail to take into account civilization's dependence upon nature's energy resources. In effect, benefiting from what can be learned from this lesson is typical of the responsibilities that accompany energy-related careers.

SCIENTIFIC INFRASTRUCTURE

If we revisit the three careers that introduced this book, another recurring theme shows through. The basic concepts and principles of science create the infrastructure upon which energy-related careers thrive. In Avery's case, the ocean thermal energy project takes advantage of precise yet amazingly simple scientific principles. Heat passes from warm to cooler materials. The absorption of heat can vaporize a liquid, but the stored energy causes no rise in temperature. Heat that is stored during vaporization is retrieved if the vapor condenses. If there is no change in pressure, the volume increases

Energy Careers: Some Common Themes 149

greatly when a liquid vaporizes, but if the container permits no increase in volume, pressures increase when liquids change to vapors.

In the OTEC process, warm seawater is pumped through heat exchangers containing liquid ammonia. The ammonia boils, and the resulting vapor creates enough pressure to drive a turbine to which is attached an electric generator. After passing through the turbine, the ammonia vapor is condensed by giving up heat to cold water pumped from approximately a thousand meters below the surface. The newly liquefied ammonia is then sent back to the evaporator to continue the energy transfer process.

Glaser's solar power satellite (SPS) system applies equally straightforward scientific principles. Light streaming from the sun actually consists of tiny bundles of energy called photons. If photons strike specially prepared surfaces, the energy they carry can be transformed to electrical energy. To prepare such a surface, thin layers of silicon that are "doped" with (contain) atoms of phosphorous are placed in close contact with a layer of silicon that is doped with boron. Energy carried by photons forces electrons to flow between the two layers of silicon. But in doing so, the energy carried by the electrons becomes available.

Rather than moving always in the same direction (direct current or DC), the flow of electrons can be made to shift back and forth (alternating current or AC) and the energy they carry can be converted to the equivalent of radio or electromagnetic waves. During the conversion process, the characteristics of the waves can be adjusted so as to minimize energy loss during passage through the atmosphere. These principles can be utilized in the reverse order. Microwave ovens, for example, begin with energy in the form of AC electricity. This form of energy, in turn, is converted to microwave, and from microwave, to heat. In fact, although the technology of solar power satellite systems is a world apart from that of a kitchen microwave oven, both owe their conception to familiarity with and the ability to utilize similar scientific principles.

Idso's career takes us one step farther. Solutions to a society's energy-related problems utilize not only the products of science but also its reasoning and investigative methods. Handling the problems that develop from burning the earth's supply of coal, oil, and natural gas certainly requires taking into account many scientific principles. It is equally necessary, according to Idso, to verify how these principles actually apply.

To claim, for example, that the biosphere will benefit, rather than suffer, from increasing the carbon dioxide content of the atmosphere requires carrying out experiments. In one such experiment, eight small sour orange trees were planted in 1987. The trees were grouped in pairs and surrounded with transparent walls of clear plastic film. Continuing to quote from Idso's article, "Cooling the Global Greenhouse?," in The World or I for March 1991, "Since November of that year, half of the enclosures have been continuously supplied with an extra three hundred parts per million (ppm) of CO_2 via a network of perforated plastic tubes lying upon the ground, while the other half have been similarly supplied with normal ambient air."

From Idso's point of view, "The effects of this near-doubling of the air's CO_2 content have been remarkable. Measurements conducted during the second and third summers of growth showed the CO_2-enriched trees to be photosynthesizing at about $2^1/_2$ times the rate of the ambient-treatment trees, while at night, the enriched trees respired or used up about 30 percent less of the daytime-sequestered carbon." But the results from the experiment are beside the point being made here. The point is that experimentation is a big part of the reasoning and investigative methods of science. Although many energy-related careers do not include working in scientific laboratories, introductory science courses provide opportunities to catch the spirit of science and to take advantage of its ways and means for bringing knowledge to bear on all kinds of personal, technological, and sociopolitical decisions.

DECISION MAKING

Energy-related decisions may involve scientific principles, but this does not free such decisions from taking into account things that can only be estimated. How much oil, gas, and coal there is in the world, for example, can only be estimated. How rapidly it will be depleted can only be predicted. Similarly, how soon nondepletable energy sources will be able to replace the fossil fuels depends upon circumstances that defy accurate predictions.

Energy-related decisions, including planning one's career, are seldom risk-free. Furthermore, not always is the most readily available information backed up by solid and unprejudiced facts. This state of affairs carries two messages. First, highly rewarding energy-related careers await people who are interested in the tactics and strategies of communication. Second, energy-related decisions depend for their accuracy upon being able to sort out and evaluate the information that becomes available.

When everything is given due consideration, energy's vital role in maintaining humanity's well-being gives energy-related careers their special appeal.

VGM CAREER BOOKS

OPPORTUNITIES IN
Available in both paperback and hardbound editions
Accounting
Acting
Advertising
Aerospace
Agriculture
Airline
Animal and Pet Care
Architecture
Automotive Service
Banking
Beauty Culture
Biological Sciences
Biotechnology
Book Publishing
Broadcasting
Building Construction Trades
Business Communication
Business Management
Cable Television
Carpentry
Chemical Engineering
Chemistry
Child Care
Chiropractic Health Care
Civil Engineering
Cleaning Service
Commercial Art and Graphic Design
Computer Aided Design and Computer Aided Mfg.
Computer Maintenance
Computer Science
Counseling & Development
Crafts
Culinary
Customer Service
Dance
Data Processing
Dental Care
Direct Marketing
Drafting
Electrical Trades
Electronic and Electrical Engineering
Electronics
Energy
Engineering
Engineering Technology
Environmental
Eye Care
Fashion
Fast Food
Federal Government
Film
Financial
Fire Protection Services
Fitness
Food Services
Foreign Language
Forestry
Gerontology
Government Service
Graphic Communications
Health and Medical
High Tech
Home Economics
Hospital Administration
Hotel & Motel Management
Human Resources Management Careers
Information Systems
Insurance
Interior Design
International Business
Journalism
Laser Technology
Law
Law Enforcement and Criminal Justice
Library and Information Science
Machine Trades
Magazine Publishing
Management
Marine & Maritime
Marketing
Materials Science
Mechanical Engineering
Medical Technology
Metalworking
Microelectronics
Military
Modeling
Music
Newspaper Publishing
Nursing
Nutrition
Occupational Therapy
Office Occupations
Opticianry
Optometry
Packaging Science
Paralegal Careers
Paramedical Careers
Part-time & Summer Jobs
Performing Arts
Petroleum
Pharmacy
Photography
Physical Therapy
Physician
Plastics
Plumbing & Pipe Fitting
Podiatric Medicine
Postal Service
Printing
Property Management
Psychiatry
Psychology
Public Health
Public Relations
Purchasing
Real Estate
Recreation and Leisure
Refrigeration and Air Conditioning
Religious Service
Restaurant
Retailing
Robotics
Sales
Sales & Marketing
Secretarial
Securities
Social Science
Social Work
Speech-Language Pathology
Sports & Athletics
Sports Medicine
State and Local Government
Teaching
Technical Communications
Telecommunications
Television and Video
Theatrical Design & Production
Transportation
Travel
Trucking
Veterinary Medicine
Visual Arts
Vocational and Technical
Warehousing
Waste Management
Welding
Word Processing
Writing
Your Own Service Business

CAREERS IN Accounting; Advertising; Business; Communications; Computers; Education; Engineering; Health Care; High Tech; Law; Marketing; Medicine; Science

CAREER DIRECTORIES
Careers Encyclopedia
Dictionary of Occupational Titles
Occupational Outlook Handbook

CAREER PLANNING
Admissions Guide to Selective Business Schools
Career Planning and Development for College Students and Recent Graduates
Careers Checklists
Careers for Animal Lovers
Careers for Bookworms
Careers for Culture Lovers
Careers for Foreign Language Aficionados
Careers for Good Samaritans
Careers for Gourmets
Careers for Nature Lovers
Careers for Numbers Crunchers
Careers for Sports Nuts
Careers for Travel Buffs
Guide to Basic Resume Writing
Handbook of Business and Management Careers
Handbook of Health Care Careers
Handbook of Scientific and Technical Careers
How to Change Your Career
How to Choose the Right Career
How to Get and Keep Your First Job
How to Get into the Right Law School
How to Get People to Do Things Your Way
How to Have a Winning Job Interview
How to Land a Better Job
How to Make the Right Career Moves
How to Market Your College Degree
How to Prepare a *Curriculum Vitae*
How to Prepare for College
How to Run Your Own Home Business
How to Succeed in Collge
How to Succeed in High School
How to Write a Winning Resume
Joyce Lain Kennedy's Career Book
Planning Your Career of Tomorrow
Planning Your College Education
Planning Your Military Career
Planning Your Young Child's Education
Resumes for Advertising Careers
Resumes for College Students & Recent Graduates
Resumes for Communications Careers
Resumes for Education Careers
Resumes for High School Graduates
Resumes for High Tech Careers
Resumes for Sales and Marketing Careers
Successful Interviewing for College Seniors

SURVIVAL GUIDES
Dropping Out or Hanging In
High School Survival Guide
College Survival Guide

VGM Career Horizons
a division of *NTC Publishing Group*
4255 West Touhy Avenue
Lincolnwood, Illinois 60646-1975